THE DYNAMICS OF THE COMPUTER INDUSTRY:
Modeling the Supply of Workstations
And their Components

THE DYNAMICS OF THE COMPUTER INDUSTRY:
Modeling the Supply of Workstations
And their Components

WALID RACHID TOUMA, Ph.D.

HD
9696
.C62
T67
1993
West

KLUWER ACADEMIC PUBLISHERS
Boston/London/Dordrecht

Distributors for North America:
Kluwer Academic Publishers
101 Philip Drive
Assinippi Park
Norwell, Massachusetts 02061 USA

Distributors for all other countries:
Kluwer Academic Publishers Group
Distribution Centre
Post Office Box 322
3300 AH Dordrecht, THE NETHERLANDS

Library of Congress Cataloging-in-Publication Data

Touma, Walid Rachid.
    The dynamics of the computer industry : modeling the supply of workstations and their components / by Walid Rachid Touma.
       p.   cm.
    Thesis (Ph. D.)--University of Texas at Austin, 1992.
    Includes bibliographical references and index.
    ISBN 0-7923-9331-7
    1. Computer industry--Mathematical models.   I. Title.
HD9696.C62T67   1993
338.4'7004--dc20   93-9419
                                                                           CIP

Copyright © 1993 by Kluwer Academic Publishers

All rights reserved. No part of this publication may be reproduced, stored in a retrieval system or transmitted in any form or by any means, mechanical, photo-copying, recording, or otherwise, without the prior written permission of the publisher, Kluwer Academic Publishers, 101 Philip Drive, Assinippi Park, Norwell, Massachusetts 02061.

Printed on acid-free paper.

Printed in the United States of America

To the **WAR CHILD**.

Contents

List of Figures   xi

List of Tables   xiii

About the Author   xv

Foreword   xvii

Preface   xix

Acknowledgments   xxi

1 Introduction   1
- 1.1 Problem Statement 1
- 1.2 Computer Technologies 1
- 1.3 Computer Systems and Workstations 2
- 1.4 Contributions 4
- 1.5 Key Findings 4
- 1.6 Outline and Summary 6

2 Definitions, Terminology, and Concepts   7
- 2.1 The Workstation 7
  - 2.1.1 The Birth of the Workstation 8
  - 2.1.2 The Workstation and Its Components 9
  - 2.1.3 The Workstation and Its Attributes 13
  - 2.1.4 Workstation Performance 13
- 2.2 Semiconductor ICs 18
  - 2.2.1 Semiconductor Physical Characteristics 19

		2.2.2	pn-Junction	20

- 2.2.2 *pn*-Junction 20
- 2.2.3 Transistor Fabrication Technologies 20
- 2.2.4 High-Speed IC Technologies 23
- 2.2.5 ICs: Speed Versus Die Area 25
- 2.2.6 Reliability of Computer Systems 26
- 2.3 Magnetic Hard Disks 26
  - 2.3.1 Computer Storage System 27
  - 2.3.2 Magnetic Storage Technology 27
  - 2.3.3 Magnetic Hard Disk Components 28
  - 2.3.4 Magnetic Hard Disk Performance 34
  - 2.3.5 Concluding Remarks 36
- 2.4 Color CRT Displays 37
  - 2.4.1 CRT History 37
  - 2.4.2 Competing Display Technologies 38
  - 2.4.3 Color CRT Components 39
  - 2.4.4 Screen Image 44
  - 2.4.5 CRT Bandwidth 45
  - 2.4.6 CRT Hardware Drivers 46
  - 2.4.7 CRT Resolution 46
  - 2.4.8 Concluding Remarks 48
- 2.5 UNIX Operating System 49
  - 2.5.1 UNIX History 49
  - 2.5.2 UNIX Design Structure 50
  - 2.5.3 Stepping Through a UNIX Command 52
  - 2.5.4 Main UNIX Managed Functions 52
  - 2.5.5 Concluding Remarks 54

3 Workstation Supply Models   **55**
- 3.1 Simulation Overview 55
  - 3.1.1 The Simulation Modeling Approach 55
  - 3.1.2 Relational Diagrams and Attributes 57
- 3.2 ICs Supply Model: Microprocessors and DRAMs 61
  - 3.2.1 Historical Data on the Physical Characteristics of ICs 61
  - 3.2.2 Historical Data on ICs Capabilities and Price Trends 70
  - 3.2.3 Model Assumptions and Terminology 78
  - 3.2.4 CPU Speed and MIPS Models 80
  - 3.2.5 DRAM Capacity Model 82
  - 3.2.6 Die Areas for Fixed Capability ICs 83

CONTENTS

	3.2.7	IC Cost Model	84
3.3	Magnetic Hard Disk Supply Model	91
	3.3.1	Historical Data on the Physical Characteristics of Magnetic Hard Disks	92
	3.3.2	Historical Data on Magnetic Hard Disk Capabilities and Price Trends	96
	3.3.3	Model Assumptions and Terminology	100
	3.3.4	Magnetic Storage Radius	102
	3.3.5	Number of Disk Recording Tracks	102
	3.3.6	Track Capacity and Density	103
	3.3.7	Hard Disk Data Rate	104
	3.3.8	Areal Capacity and Density	106
	3.3.9	Volumetric Capacity and Density	106
	3.3.10	Magnetic Hard Disk Cost per Megabyte	106
	3.3.11	Magnetic Hard Disk Total Cost	109
3.4	Color CRT Display Supply Model	109
	3.4.1	Historical Data on the Physical Characteristics of Color CRT Displays	110
	3.4.2	Historical Data on Color CRT Display Capabilities and Price Trends	113
	3.4.3	Model Assumptions and Terminology	115
	3.4.4	Bandwidth	117
	3.4.5	Resolution	117
	3.4.6	Metal Shadow Mask Manufacturing Yield	120
	3.4.7	Color CRT Display Cost	121
3.5	UNIX Operating System Supply Model	124
	3.5.1	UNIX Development-from-Scratch and Porting Trends	124
	3.5.2	Model Assumptions and Terminology	126
	3.5.3	UNIX Development-from-Scratch Time Period	128
	3.5.4	UNIX Porting Time Period	129
	3.5.5	Software Attributes Index	130
	3.5.6	Workstation Hardware Attributes Index	130
	3.5.7	UNIX Development-from-Scratch and Porting Costs	132
3.6	Workstation Assembly Model	133
	3.6.1	Assembly Steps	136
	3.6.2	Model Assumptions and Terminology	136
	3.6.3	Model Formulation	137

4 Model Behavior and Sensitivity Results — 141
- 4.1 Component Cost, Single Unit Price, and Bulk Price ... 143
- 4.2 Component Supply Model Inputs 143
  - 4.2.1 Model Input Parameters 144
  - 4.2.2 Components' Physical Characteristics Trends ... 145
  - 4.2.3 Components' Capabilities and Price Trends 145
- 4.3 ICs: Model Results and Actual Market Data 146
  - 4.3.1 ICs Die Yields 146
  - 4.3.2 ICs: CISC CPUs Model Results and Actual Market Data........................ 147
  - 4.3.3 ICs: RISC CPUs Model Results and Actual Market Data........................ 149
  - 4.3.4 ICs: DRAMs Model Results and Actual Market Data........................ 151
- 4.4 Magnetic Storage: Model Results and Actual Market Data 156
  - 4.4.1 Magnetic Hard Disk Price per Megabyte 156
  - 4.4.2 Magnetic Hard Disk Areal Density 156
  - 4.4.3 Notes on Volumetric Density and Data Rate ... 158
- 4.5 Color CRT Display: Model Results and Actual Market Data 159
  - 4.5.1 Color CRT Price per Megapixel 159
  - 4.5.2 Color CRT Number of Pixels per Inch 160
- 4.6 UNIX: Model Results and Actual Market Data 161
  - 4.6.1 UNIX Porting Times and Costs 161
  - 4.6.2 UNIX Development-from-Scratch Time and Cost . 161
- 4.7 Workstation Assembly Model: Inputs and Projected Results 164
  - 4.7.1 Projected Results 165
- 4.8 Sensitivity Analyses 167
  - 4.8.1 Sensitivity to the Feature Size 169
  - 4.8.2 Sensitivity to the Number of Silicon Wafer Defects per Unit Area...................... 181

5 Conclusions and Suggestions for Future Research — 187
- 5.1 Suggestions for Future Research 189
- 5.2 Final Comments 190

References — 193

Index — 203

List of Figures

1.1	Computer classes versus user environment.	3
2.1	General computer system configuration.	10
2.2	*npn* bipolar transistor configuration.	21
2.3	Enhancement type NMOS transistor configuration.	22
2.4	CMOS inverter configuration.	23
2.5	Rigid disk file components.	29
2.6	Magnetic disk surface layout.	31
2.7	Inductive head structure.	32
2.8	CRT display components.	39
2.9	Inner screen phosphors layout structures.	43
2.10	UNIX layer structure.	51
3.1	Illustration of a relational diagram.	57
3.2	Die sizes of CISC CPUs and DRAMs versus time.	63
3.3	Feature size versus time.	66
3.4	Silicon wafer diameter versus time.	69
3.5	Actual data on the number of instructions per cycle for Intel and Motorola CISC CPUs.	75
3.6	Actual data on the number of instructions per cycle for HP-PA and Sun SPARC RISC CPUs.	76
3.7	Relational diagram of the operational speed of CPUs.	80
3.8	Relational diagram of the DRAM capacity.	83
3.9	Relational diagram of the average IC die testing time.	88
3.10	DASD recording system scaling.	93
3.11	Actual price per megabyte of magnetic hard disk storage versus time.	97

3.12 Areal densities of magnetic storage devices in bits per mm^2 versus time. 99
3.13 Illustration of the dimensional parameters of a magnetic hard disk. 103
3.14 Relational diagram of the cost per megabyte of a magnetic hard disk. 107
3.15 Relational diagram of the number of pixels per inch of a color CRT display. 118
3.16 Relational diagram of the cost per megapixel of a color CRT display. 122
3.17 Relational diagram of the UNIX development-from-scratch time period. 128
3.18 Relational diagram of the UNIX porting time period. . . . 130
3.19 Relational diagram of the workstation hardware attributes. 132
3.20 Illustration of a workstation assembly network. 135

4.1 Present value prices of workstations: types 1, 2, and 3. . . 166
4.2 Feature size: sensitivity of the price per megaHertz of CPUs — cases 1, 2, and 3. 171
4.3 Feature size: sensitivity of the price per megabyte of DRAMs — cases 1, 2, and 3. 173
4.4 Feature size: sensitivity of the price per megabyte of magnetic hard disks — cases 1, 2, and 3. 175
4.5 Feature size: sensitivity of the price of a 19-inch color CRT display — cases 1, 2, and 3. 176
4.6 Feature size: sensitivity of the present value price of a type 2 workstation — cases 1, 2, and 3. 178
4.7 Feature Size: Sensitivity of the DRAM and the magnetic hard disk prices per megabyte — cases 1 and 3. 180
4.8 DPUA: sensitivity of the price per megaHertz of CPUs — cases 1, 2, and 3. 184
4.9 DPUA: sensitivity of the price per megabyte of DRAMs — cases 1, 2, and 3. 185

List of Tables

3.1	Die areas and their actual die yields.	65
3.2	Actual market data of Intel and Motorola CISC CPUs.	71
3.3	Actual prices and price per MIPS market data of Intel CISC CPUs.	72
3.4	Actual performance data of HP-PA and Sun SPARC RISC machines.	74
3.5	Actual market data on DRAM die sizes, capacities, and densities.	77
3.6	Actual prices and price per megabyte market data of Motorola DRAMs.	78
3.7	Actual market data of 19- and 20-inch color CRT displays.	112
3.8	Actual price per megapixel and number of pixels per inch market data of 19- and 20-inch color CRT displays.	115
4.1	Model results and actual market data of die yields for certain die areas in 1989.	146
4.2	Model results and actual market data on MIPS ratings and prices of Intel CISC CPUs.	148
4.3	Model results and actual market data on the price per MIPS of Intel CISC CPUs.	149
4.4	Model results and actual market data on the speed per cm^2 rating of CISC CPUs.	150
4.5	Model results and actual market data on MIPS ratings and prices of HP RISC CPUs.	152
4.6	Model results and actual market data on the speed per cm^2 rating of RISC CPUs.	152
4.7	Model results and actual Motorola market data on DRAM capacities and prices.	153

4.8	Model results and actual Motorola market data on the price per megabyte of DRAMs.	154
4.9	Model results and actual Motorola market data on the number of DRAM megabytes per cm^2.	155
4.10	Model results and actual market data on the price per megabyte of magnetic hard disks.	157
4.11	Model results and actual market data on the areal density of magnetic hard disks.	158
4.12	Model results and actual market data on the price per megapixel of a 19-inch color CRT display.	159
4.13	Model results and actual market data on the number of pixels per inch of a color CRT display.	160
4.14	Model results and actual market data on the UNIX porting time periods and costs.	162
4.15	Model results and actual market data on the UNIX development-from-scratch time periods and costs.	163
4.16	Projected prices of the workstation assembly components and raw materials.	165
4.17	DPUA: sensitivity of the die yields for certain die areas — cases 1, 2, and 3.	182

About the Author

Walid Rachid Touma was born in Kab-Elias, the Bekaa Valley, Lebanon, on April 2, 1965. In 1984, he joined the engineering program of the University of Texas at Austin. He was honored as an Engineering Scholar and received the degree of Bachelor of Science, with High Honors, in Electrical and Computer Engineering in May 1987. In September 1987, he joined the Graduate School of the University of Texas at Austin and earned the degree of Master of Science in Computer Engineering in December 1989. His thesis was "Clustering/Partitioning Algorithms and Comparative Analysis." In January 1990, he assisted his supervisor, Professor Martin L. Baughman, and five other faculty at the University of Texas at Austin in obtaining a grant from DARPA to fund the "Workstation Project." In May 1992, he earned his Ph.D. in Computer Engineering. His dissertation, "The Dynamics of the Computer Industry: Modeling the Supply of Workstations and their Components," was nominated for the Outstanding Dissertation Award at the Department of Electrical and Computer Engineering.

Currently, he is developing an interactive software tool that complements the models presented in this book.

Foreword

The pace of change in the computer industry is fueled by rapidly increasing microelectronics manufacturing capabilities, growing demands for ever more sophisticated user applications, and a very competitive business of assembly and marketing. The computer workstation, which incorporates the high end of the industry's technical capabilities into a network interface with impressive computational and graphics powers, is the technology of tomorrow.

There are several reasons why this book will be of interest to the reader. First, the topic is very innovative, the work is original, and its publication is timely. The workstation industry was in its infancy in the decade of the 1970s, its adolescence in the 1980s, and moving into young adulthood in the 1990s. This book covers all three time frames. What the author has done is combine analysis of the engineering trends in computer workstations with information about the economics of manufacturing the workstation components to produce a unique view of the dynamic behavior of this industry—past, present, and future. To my knowledge this work is path breaking.

A second reason this book will be of interest to the reader is that it is very well written in a format and style of presentation that makes it easy to read. Computer engineering is a highly specialized technical field. Yet, I believe that non-specialists will find the work reported in this book accessible, while specialists will appreciate the breadth and depth of the presentation of the engineering and economic trends analyzed.

Finally, and the third reason this book will be of interest to the reader, is that the insights provided by this work into the dynamics of the supply of the workstation components, both in terms of direction and pace of change, are stunning. The collection of models presented and the results produced represent impressive research products. They not only illuminate the important drivers of technical change in this industry, but they also offer future analysts and students of this industry a documented standard for further analysis and modeling endeavors.

This is the first; there will surely be many to follow.

Martin L. Baughman
Professor
Department of Electrical and Computer Engineering
The University of Texas at Austin
Austin, Texas

Preface

Salvador Dali writes:

> At three, I wanted to be a cook. At five, I wanted to be Napoleon. My ambition has been growing ever since, and now my ambition is to become Salvador Dali, nothing else. It is, nevertheless, very difficult, because the closer I come to Salvador Dali, the farther away from me he goes.

As you are reading these words, the computer industry is already in a new metamorphosis, a new cycle, a new era.

Such dynamics intrigued me.

It made me wonder and question how obsolete my knowledge is and what type of cycles I ought to follow to keep up with the industry's roller coaster-like dynamics.

I am sure one might have to spend a lifetime trying to understand the self; incidently, with the way the computer industry is mutating, I am sure a lifetime might not be enough.

Nevertheless, whether you are an artist/scientist, an engineer, or an economist/analyst, I know you will find something in this book that will make you question what you already know, and what you should attempt at knowing better.

Enjoy.

Acknowledgments

Father Rachid, I love you and I respect you. You are my soul friend, you are my mentor in life and my inspiration, and I would not be who I am today without your emotional, mental, and moral support. Mother Laure, I love you and I kneel to the effort you have put into rearing me. I only wish I could have spent more time with you in my adult life so that I could have gotten to know you better. I guess the war has its prices.

Marty Baughman, I thank you and I sincerely acknowledge your mentorship and your partnership. We met in 1985 and, since then, we have had a genuine friendship, coupled with mutual respect and appreciation for each other's talents. Our friendship came to life as we teamed up on the "Workstation Project" and developed what has become, so far, a new research area that deals with understanding the dynamics of one of the most fascinating industries of our time: the computer industry. As our endeavors progressed, several new members joined the team, such as Jacob Abraham, David Kendrick, Joe Rahmeh, Steve Szygenda, and Sten Thore, my acknowledgements of whom will be presented later.

Katherine Gregory, my soul mate, my music, my metamorphosis, my woman: you have been the pillar of emotional support for me while I wrote this book. Here's to you, Kathy G., to the wonderful, loving, and innocent child in you, to the mother in you, and to your soul.

Brothers Tanos and Jihad and your families, and Sister Nadine, my best friends, my ears, my partners, and my support, I love you. Uncle Fadlo, my wise ear away from home, I love you, I sincerely appreciate the advice you have given me, and I thank you for assisting me in coming to the United States and in joining the University of Texas at Austin.

Jacob Abraham, I appreciate your support for the project since its inception. David Kendrick, I thank you for your wisdom, your honesty, and your patience. Joe Rahmeh, I thank you for all your support throughout my graduate career. Steve Szygenda, I thank you for your constructive criticism throughout the project. Sten Thore, at times, you have been like a father to me, and I am deeply grateful to your generosity and genuine soul.

Last but not least, I would like to thank everybody at the departments of Electrical and Computer Engineering, Economics, Mathematics, and Business, at the Computer Engineering Research Center, at the Center for Economic Research, and at the IC^2 Institute who assisted the team and me throughout the "Workstation Project" and while I wrote this book.

Walid Rachid Touma

1

Introduction

1.1 Problem Statement

The digital information age is here. Computers across the globe communicate via satellite or fiber optic links, wide area networks share resources thousands of miles away, and a child at home has access, at the press of a button, to a world of knowledge. Several technologies have made possible this computer era, driven it, and affected its dynamics over time. The problem to be addressed in this book is the formulation of a model that interrelates the factors that drive the supply of these technologies over time to the attributes of the computers that are manufactured from them.

1.2 Computer Technologies

A computer system is a grouping of resources, hardware and software, accessed by application programs that conform to the computer's programming language. The hardware resources are the seven component assemblies of the computer system: the processing board, the memory board, the data storage system, the display monitor, the printer, the mouse, and the keyboard. Each hardware resource is in itself a collection of several components, the functions of which will be elaborated upon later. The software resources consist of two major components: the computer system's manager, referred to as the operating system, and the applications programs.

The attributes of the above technologies, at any point in time, depend on a number of factors, dictated largely by the tradeoffs between product life, production yields, learning curves, the pace of technical change, competition, and the state of technology. These factors are interrelated here in deterministic simulation models. The behavior of each resource's supply is dynamically represented and then integrated with dynamic models of other resources to provide insight into the dynamics of the assembled computer products. The component supply models, to be presented in a later chapter, include an integrated circuits (ICs) supply model: microprocessors and dynamic random access memories (DRAMs), a magnetic hard disk storage supply model, a color cathode-ray tube (CRT) display supply model, and a UNIX operating system supply model.

1.3 Computer Systems and Workstations

Every computer system, once configured, has its own application-specific attributes and its own market niche. Figure 1.1[1] illustrates this concept by relating the computer user environment on the horizontal axis to the computer use style on the vertical axis, and ranks the available computer systems with respect to those two criteria. For instance, the computer workstation is represented in this figure as a shared computer in a professional environment; alternately, the personal computer (PC) is represented as a personal and private computer in a professional environment.

In the remainder of this book, the computer workstation is focused upon because it combines the latest developments in processing power[2], DRAM capacities and densities, data storage system capacities and data rates, display resolutions, and computer management and computer applications software. Workstations, when interconnected to one another, form a single, shared (work and files) but distributed computing environment [4]. This concept of network computing and communication has catapulted the sales growth of the workstation higher than any technology in the computer industry, and sales are projected to grow at a compound rate of 30 percent per year over the next five years [86]. Due

---

[1] Adele Goldberg, **A History of Personal Workstations**, ©1988 ACM Press, a division of the Association for Computing Machinery, Inc. Reprinted with permission of Addison-Wesley Publishing Company, Inc.

[2] In this context, power is equivalent to speed and functionality.

1.3. COMPUTER SYSTEMS AND WORKSTATIONS

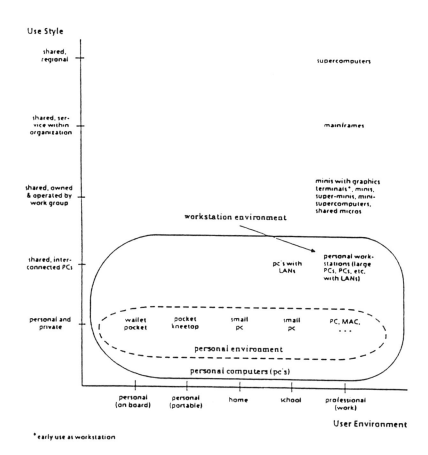

Figure 1.1: Computer classes versus user environment. Source: [4]. ©1988 ACM Press.

also to its high performance to price ratios [59], the computer workstation has emerged as the driving force in computing, a role formerly filled by mainframes and supercomputers.

1.4 Contributions

During the past four decades, the computer industry has emerged as one of the most—if not the most—dynamic industries since the industrial revolution. Never in history has there been a product with so short a lifetime, or period to obsolescence, compared to the time and effort that go into its research, development, and demonstration. One of the main objectives of this book is to shed some light on the technology-driving trends of the computer industry and their effects on the dynamics of the industry as a whole, and in particular, on the supply of assembled workstations and their components.

The contributions of this book to the state of knowledge are:

- Detailed analyses and documentation of the historical trends in the supply of hardware and software components used in a computer workstation.

- Deterministic discrete event simulation models that capture the past dynamics of various attributes of workstation components.

- A workstation assembly model that brings together all the supply models of the workstation components to provide insight into possible future behavior of the supply of fixed capabilities workstations.

- A tool to:
  - project the trends in the supply behavior of the industry for alternative scenarios of market or technological change.
  - perform sensitivity analyses with respect to certain technology barriers and their effect on the decision-making strategies of the companies supplying the technologies.

1.5 Key Findings

The key findings of this book are as follows.

1.5. KEY FINDINGS

1. A $10,000 price ceiling has become a standard for top-of-the-line workstations in today's distributed computing environment. The model results from this book show that by 1994, a top-of-the-line workstation will have the following hardware capabilities: a 200 megaHertz CPU, 64 megabytes of DRAM, a 1 gigabyte of magnetic storage, and a 19-inch color CRT display with a 2.6 megapixel resolution. These same capabilities, if configured into a workstation in 1991, would cost approximately $20,000. This is a rate of decrease in price of over 20 percent per year for a fixed capabilities workstation.

2. The feature size of integrated circuits is one of the most critical and influential technology-driving trends in the computer industry. The price results of all the workstation component supply models are sensitive to changes in the rate of decrease of the feature size over time. The effects of these changes are reflected in the overall price of an assembled workstation. In the base case, the feature size decreases at an exponential rate of 5.5 percent per year. In case 2, the feature size stops decreasing after 1992, and the price of an assembled workstation jumps 20.5 percent from the projected base case price for 1996. In case 3, the feature size decreases at an exponential rate of 10 percent per year—almost double the base case rate—and the price of an assembled workstation decreases 8.4 percent from the projected base case price for 1996.

3. One of the possible consequences of accelerating the rate of decrease of the feature size is an accelerated decrease in the price per megabyte of semiconductor DRAMs. When the rate of decrease of the feature size doubles after 1992, the model results show that the price per megabyte of a semiconductor DRAM becomes cheaper than the price per megabyte of a magnetic hard disk by the year 2001, assuming also a continued fall in magnetic disk drive prices. If nonvolatile semiconductor DRAMs are developed by 2001, DRAMs could change the overall configuration of a computer by replacing the magnetic hard disk as the permanent storage component and pave the way to real-time computing.

4. A clean and precise ICs manufacturing environment and, consequently, a smaller number of silicon wafer defects per unit area greatly influence the reliability of the chips, their yields, and, ulti-

mately, their costs. In the base case, the number of silicon wafer defects per unit area is 2.5 defects per cm^2. In case 2, the number of silicon wafer defects per unit area doubles after 1992 to 5 defects per cm^2, the IC die yields from the model decrease by as much as 88.9 percent, and the CPU price per megaHertz and the DRAM price per megabyte jump 277 percent and 69.1 percent, respectively, from their projected base case values for 1996. In case 3, the number of silicon wafer defects per unit area is halved to 1.25 defects per cm^2 after 1992, the IC die yields increase by as much as 355.6 percent, and the CPU price per megaHertz and the DRAM price per megabyte decrease 51.8 percent and 47.7 percent, respectively, from their projected base case values for 1996.

1.6 Outline and Summary

This book is organized as follows:

- Chapter 2 presents some of the main definitions, terminology, and concepts related to computer workstation technologies.

- Chapter 3 presents discrete event simulation supply models of microprocessors, DRAMs, magnetic hard disks, color CRT displays, and UNIX operating systems, and a linear workstation assembly process model.

- Chapter 4 presents simulation results from the models. It compares the results of the component supply models with historical trends, presents the results of the workstation assembly model for three different workstation configurations, and analyzes the sensitivity of the component supply and workstation assembly models to variations in the projected rate of decrease of the IC feature size and in the number of silicon wafer defects per unit area.

- Chapter 5 presents the conclusions and suggests future research directions.

2

Definitions, Terminology, and Concepts

This book relies upon concepts relating to both computer industry technologies and mathematical modeling. These concepts are merged to create models that capture the dynamics of the supply of workstations and their components. This chapter provides some of the essential definitions, terminology, and concepts related to computer technologies. The mathematical modeling concepts will be discussed in detail in a later chapter.

Section 2.1 presents an overview of the workstation and its attributes. Section 2.2 describes the manufacturing processes and attributes of semiconductor ICs. Section 2.3 describes the assembly and attributes of magnetic hard disks. Section 2.4 describes the assembly and attributes of color CRT displays. Section 2.5 presents an overview of the design and functionality of the UNIX operating system.

Volumes could be written about each of these technologies and its attributes. Only the basics are presented here. References are provided for readers interested in more complete descriptions.

2.1 The Workstation

Since the workstation is the computer system under consideration here, the following subsection presents a brief history of the emergence of the workstation. In subsection 2.1.2, a description of the workstation component assemblies is presented. In subsection 2.1.3 a workstation attribute is defined, and in subsection 2.1.4 the factors affecting the workstation's performance and how to measure them are reported.

2.1.1 The Birth of the Workstation

The computer industry can be compartmentalized into user applications and computer system attributes that match these applications. The system attributes depend on the capabilities and specifications of the system's hardware and software components. Such capabilities will differentiate the system from other computers in price and performance. From the point of view of the user familiar with personal computers (PCs), the workstation is an advanced PC; i.e., a workstation has more processing power, more main memory, more data storage space, a better user interface, and more sophisticated display capabilities than a PC. Workstations edged out other forms of computing when they were introduced in 1980 with their networking capabilities, which are handled by the operating system or the software manager of the machine [4, 29]. Chief among the technical developments that led to the birth of the workstation were:

- The development of an operating system that handled distributed and multitasking computing environments. The Digital Equipment Corporation (DEC) and Xerox were the leading developers of such software [4].

- The commercialization of the local area network (LAN) software and the Ethernet hardware in 1981.

- Breakthroughs in the semiconductor industry, especially the availability of cheaper and denser memory chips and a more powerful array of microprocessors and microcontrollers.

- The increase in the disk[1] storage density and its data-access rates.

- Availability of displays with over 1 million pixels, driven by sophisticated hardware that enabled higher refresh rates and better resolutions.

---

[1] A disk, in this context, is a data storage system that uses either magnetic or optical technologies.

2.1.2 The Workstation and Its Components

All workstations have a common set of component technologies. Illustrated in Figure 2.1 are the three major hardware component assemblies and the software component assembly. The hardware component assemblies include the processing board, the memory board, and the external input/output (I/O) interfaces (keyboard, mouse, data storage, display, and printer). The software component assembly includes the system resources manager—operating system—and the computer/user window-interface manager[2]. Each hardware component assembly is shown in a bold-typed rectangle and connected to another hardware assembly or assemblies by a bold-typed line. The software component assembly is shown in a broken rectangle to indicate that it is not a hardware component, but acts as a hardware component assemblies manager.

Some components in the hardware assemblies are shown in broken circles to illustrate that their presence is optional, like the cache[3] in the processing board assembly. In the case of the storage technologies, the magnetic and the optical boxes are connected to the storage element of Figure 2.1 with broken lines to indicate that either could be used as the storage technology in the workstation. Similarly, the CRT and the LCD[4] boxes are connected to the display element of Figure 2.1 to indicate that either could be used as the display technology in the workstation.

The hardware component assemblies are connected in a hierarchical order: first, the processing and the memory boards where the task execution is performed; and second, the I/O interfaces where the task can be keyed in, stored, displayed, or printed. The software component is shown in Figure 2.1 between the two hardware layers of the computer system to indicate that it manages the execution of the task and its interaction with the I/O interfaces. The two hardware layers are connected by communication lines, the collection of which is called the bus. A bus can be 8 bits[5] to 128 bits wide, depending on the architectural configuration of

---

[2]The system resources manager and the computer/user window-interface manager can reside permanently in a storage device (magnetic or optical hard disk), or temporarily in main memory while the computer system is powered on.

[3]If the machine price is not an issue, optional cache memories can be installed on the main memory board and on hardware driver boards of the I/O interfaces.

[4]LCD is equivalent to liquid crystal display.

[5]The bit is the digital representation of information. Its values are binary (0 or 1), and a series of eight bits represents a byte.

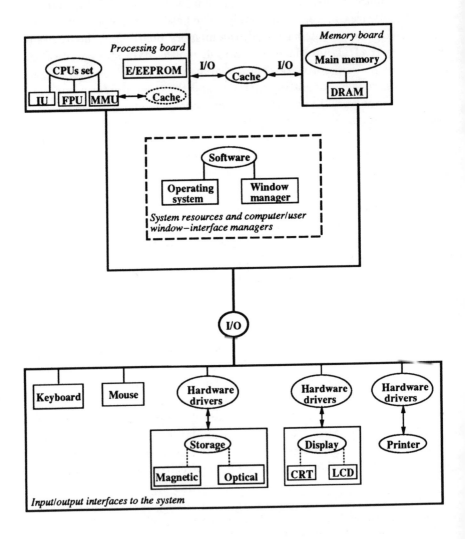

Figure 2.1: General computer system configuration.

2.1. THE WORKSTATION

the microprocessor, the main memory[6], and the I/O interfaces.

What follows is a brief description of the component assemblies in Figure 2.1 and their corresponding subcomponents:

- The processing board includes an integer unit, a floating point unit, a memory management unit, and, in some cases, the cache memory and/or read-only-memory (ROM) chips. The integer and floating point units handle several types of instructions such as arithmetic, logical, control, floating point, or data transfer operations. The memory management unit handles the interfaces between the processing units and the cache memory, between the cache memory and the main memory, and between the main memory and the disk memory. The cache is a static random access memory (SRAM) sandwiched between the processing units and the main memory to reduce the wait-on-data cycles of the processors. Usually, the cache is the fastest type of memory used in the workstation and has the highest cost per byte of memory used. The ROMs[7] contain subroutines that perform specific tasks such as computer startup and initialization procedures.

- The main memory board consists of DRAMs. The application program and the data it manipulates are located in DRAMs during the program's execution. An application program is a sequence of instructions that perform a task. An instruction specifies the arithmetic and logic operations to be executed and governs the transfer of information within the computer's resources. Data in computers are any digital information representing numbers and encoded characters to be used during the program execution; data can be entered in the computer system via the keyboard or loaded into main memory via a backup storage device, such as a magnetic hard disk, tape, or diskette. Once the program is in main memory and ready to be executed, the processor fetches the instructions and performs the desired operations. When the program finishes execution, the results are either displayed on the monitor or stored onto disk for later reference. For a more elaborate description of

---

[6]Main memory can be considered as an I/O interface to the microprocessor, but in the hierarchy adopted, it is a part of the processing team.

[7]ROMs are usually configured as electrically erasable and programmable read-only-memories (EEPROMs).

the process of executing a program, see Hamacher, Vranezic, and Zaky [31], and Hennessy and Patterson [33].

- The data storage system consists of magnetic or optical disks. These disks retain the digitized information indefinitely, unlike the DRAMs where the data disappear as soon as the machine's power is turned off. Disk data rates (megabytes per second) and areal densities (bits per inch^2) affect the efficiency of the computer system and the types of applications it can perform. The communication between the hard disk and the other computer resources is the most time-consuming operation and, to reduce the resulting communication delays, several cache systems have been implemented between the disk and the computer's main memory. If the data requested by a program in execution are not in main memory, the processor has to stall[8], and the execution is stopped until the disk comes back with the data. Hence, a trend in computer systems design will provide enough main memory[9] in the machine so that the program and most of its data are loaded in main memory before the execution starts, minimizing—if not eliminating—the references to the disk during execution.

- The high-definition display consists of a color or black and white (B&W) cathode-ray tube (CRT) or liquid crystal display (LCD), and the necessary image-driving hardware. The image drivers are fast input/output (I/O) chips and play an important role in determining the display's resolution, refresh rates, and color templates. To the user, the display in a computer system is the main tool for displaying the output of an application program. Printers are also output devices, but they are too slow, expensive, and impractical for the highly interactive and animated software available on the market today; their best use is in obtaining a hard copy of finished work.

- For the past decade, UNIX has been the most effective operating system for workstations. UNIX manages the workstation's resources and coordinates the interface between the user and the machine and between the user and other network workstations. Of

---

[8] In multitasking machines, the processor could perform another task while waiting for the data from the disk.
[9] Assuming that dense and cheap DRAMs are available.

utmost importance to the workstation environment is its ability to network with other machines. This capability is supported by the UNIX-distributed environment which provides the platform and the communication protocols for building large and wide area networks (LANs and WANs) of workstations. Several "window"-based software applications have been developed in the 1980s to manage and facilitate the computer/user interface. One of the most popular "window"-based systems running on top of UNIX in computer workstations is the X windows manager[10].

2.1.3 The Workstation and Its Attributes

A workstation attribute is a measurable characteristic that enhances the workstation's value and reflects the sophistication of its technology. Each of the workstation components presented in subsection 2.1.2 has its own specifications and adds enhancements to the operation of the workstation as a whole. Of the many workstation attributes, the main ones are the processing board's speed, the amount of DRAM, the storage system's capacity and response time, the display's resolution and color, the operating system's user friendliness, networking, and multitasking capabilities, and, most importantly, the workstation's total cost. All of the previously mentioned attributes[11] come together to influence the workstation's overall performance attribute, presented in detail in the next subsection.

2.1.4 Workstation Performance

The performance of a workstation is measured by the time a program takes to finish its execution. The execution time is measured by the following equation:

$$Time_{exec} = IPP * CPI_{ave} * CT \ (sec) \qquad (2.1)$$

---

[10]The X windows manager was developed at the Massachusetts Institute of Technology (MIT) with the support of the Digital Equipment Corporation (DEC); currently, an X consortium is handling the system's updates and enhancements.

[11]Attributes such as workstation size and weight are not discussed here because they require the study of workstation demand attributes, and this book is concerned with supply attributes only.

where *IPP* is the number of instructions per program, CPI_{ave} is the average number of machine clock[12] cycles per instruction, and CT is the clock cycle time in seconds. The designer of a high performance machine reduces the average execution time of a program by effectively reducing the factors contained in the right-hand side of equation (2.1). The elements affecting the right-hand side factors of equation (2.1) and a description of the workstation performance measurement techniques are presented below.

IPP: The Number of Instructions per Program

The number of instructions per program depends on the compiler and the architecture of the machine. A compiler maps the source code written in a high-level language, such as C or Fortran, onto the machine's hardware language or assembly code. The most common machine architectures are CISC[13] and RISC[14].

The **CISC** instruction set is complex because most instructions require several minor operations during execution. The instructions have varying lengths in bits[15], making the decoding[16] circuitry and the memory access procedures more elaborate and complicated. A CISC instruction, on average, takes more than one cycle to terminate; nevertheless, the number of CISC instructions per compiled program is small compared to a RISC compilation.

The initial objective of **RISC** architecture was to have a set of simple one-cycle instructions [15]. Several scientists and institutions worked on the development of the RISC architecture, pioneered by IBM with the 801 minicomputer project [15], by David A. Patterson with the RISC[17] project at the University of California at Berkeley, and by John L. Hennessy with the MIPS project at Stanford University. All RISC instructions have the same size—32 bits; most of them require one cycle to terminate, except floating point instructions which require more

---

[12]The machine clock is the computer cycle time or the inverse of the frequency at which the computer is running.

[13]CISC is an acronym for Complex Instruction Set Computing.

[14]RISC is an acronym for Reduced Instruction Set Computing.

[15]In a digital machine, data and instructions are represented by bits.

[16]Decoding is performed on a fetched instruction from main memory into the processor to determine the procedures the processor must follow to execute the fetched instruction.

[17]The first RISC microprocessor, RISC 1, was designed in 1982.

2.1. THE WORKSTATION

than one cycle. Complex floating point operations are usually software-implemented or rerouted to a dedicated floating point processor. Memory access instructions in RISC are very simple, composed of basic load and store commands. Such simplicity in the instructions set induces a simpler hardware design than CISC's and, consequently, faster machine operational speeds. Even though the RISC instruction set is simple, it takes several RISC instructions to perform comparably to one CISC instruction. There are tradeoffs, then, with both architectures: a CISC program has fewer instructions than a RISC, and a RISC instruction requires fewer computer cycles to terminate than does a CISC.

Once the instruction set is chosen, an implementation technique for reducing the number of instructions per program is to optimize the *compiler*'s source code[18] to object code[19] mapping techniques [15, 34]. Compiler technologies flourished with the availability of cheap and dense DRAMs and the increase in the demand for RISC-based workstations. When main memory became less expensive, machines with greater amounts of DRAMs were affordable, and larger programs with more instructions could reside in them. And, since a RISC compiled program has a great number of instructions, efficient mapping techniques became the focus of RISC compiler developers to compete with the speed of execution of a CISC. RISC compilers were designed to optimize the distribution of the code within the constraints of the computer's architecture, even if the architecture was not the most suitable for the application [83].

CPI_{ave}: The Average Number of Clock Cycles per Instruction

Several hardware organization and implementation factors influence a computer's average number of clock cycles per instruction. The most important ones are instruction pipelining and memory caching [34]. Instruction pipelining[20] is an architectural paradigm that improves the throughput of the machine without changing the basic cycle time by paralleling instruction executions in one processor. A pipeline can have several stages, where each stage gets dedicated to an instruction until it terminates. Since each instruction has several steps to be performed before termination, pipelining those steps can result in, on average, one

---

[18] Source code is equivalent to high level code like C.
[19] Object code is equivalent to machine language or assembly code.
[20] Instruction pipelining was introduced in mainframes in the early 1960s.

instruction per computer clock cycle and, in some cases, two or more [72, 99], depending on how many instructions are launched simultaneously per cycle and on the degree of sophistication of the hardware implementation.

The throughput can be increased by running in parallel more than one *functional unit*, with each unit performing a particular task[21]. In today's workstations, this form of parallelization is sometimes implemented through one dedicated integer unit (IU), one floating point unit (FPU), and one memory management unit (MMU) per machine. These units may be integrated in one chip or separate, and most of their operations are hardware controlled. The form of parallelization that integrates more than one similar processing unit into one workstation[22]—that is, the machine is configured with several IUs, FPUs, and MMUs—has been developed [87], but cost reduction and software improvements are needed before such workstations gain any market share.

Memory caching is the most vital operation between the processor and the workstation's resources. Implemented inside or outside the processor, caches are the key tool in reducing the wait-on-data time (from main memory or disk) of a program during execution. Moreover, since the DRAM access speed increases at a much lower rate than do processors [34], more cache[23] memories are needed to fill that communication speed gap to accommodate the much faster operating rates of the processors.

CT: The Clock Cycle Time

The clock cycle time is the inverse of the hardware operating frequency (clock speed). Several workstations today operate at frequencies larger than 50 megaHertz (MHz), like the Hewlett-Packard (HP) 9000/730 (RISC) and the Intel *i*486 based machines (CISC) which operate at 66 megaHertz. It is safe to assume that speeds in the 200 to 250 megaHertz range will be attained before the end of the decade [65, 72, 99]. Nevertheless, several physical barriers may become apparent when speeds reach

---

[21] The superscalar architecture of the IBM RISC System/6000 workstation implements this idea.

[22] A computer with more than one similar processing unit is called a parallel machine. Most parallel machines have either a SIMD (single instruction-multiple data) or a MIMD (multiple instructions-multiple data) configuration.

[23] Since the costs of cache memories and their controllers are decreasing, caches have found their way into hard disk controllers and display and printer drivers.

2.1. THE WORKSTATION

the 0.5 to 1 gigaHertz (GHz) range, among them the wave reflection phenomenon. As the distance traveled by the electric signal decreases to accommodate higher frequencies, problems with wave reflections occur [83]. A wave reflection occurs when a signal generated by the line driver is received by a circuit whose impedance[24] is less than the line's impedance. To preserve Ohm's law[25], part of the signal travels back in the direction of the driver, and a wave is reflected; the noise generated by these reflections creates erroneous behavior in the whole machine. Other physical barriers are the fabrication of the ICs and their packaging. These will be described in later sections.

Workstation Performance Measurement

Measuring the performance of a workstation is an art [33, 44, 53, 100]. Using benchmark programs is the most common procedure to measure the performance of a workstation. SPEC[26] is the leading benchmark currently in use in the computer industry [100]. It is a collection of ten programs—four integer intensive and six floating point intensive programs. The performance of a machine is expressed as the geometric mean of the respective ratios of the execution time of the ten benchmark programs on the VAX 11/780 to their execution time on the benchmarked machine. Since most of the other popular benchmarks do not consider the system's configuration[27], load distribution, or the size of the tasks run, the system's evaluator must incorporate their effects into the overall performance results before reporting them [44].

Metrics performance measures, like MIPS[28] or MFLOPS[29], have gained popularity since the emergence of RISC architectures, and they have been used to lure CISC users to RISC with the higher performance metrics obtained by RISC-based workstations. Even though the MIPS and MFLOPS metrics do not report on the overall system throughput or re-

---

[24]The impedance is equal to the ratio of the phasor equivalent of a steady-state sine-wave voltage to the phasor equivalent of a steady-state sine-wave current, V/I.

[25]Z_{ohms} = impedance = V/I.

[26]SPEC is an acronym for Systems Performance Evaluation Cooperative.

[27]The configuration of the system is equivalent to the chosen processors and their speeds, the amount of DRAM, the capacity of the hard disk, and the type of operating system: each MIPS ought to be matched with 1 megabyte of DRAM and 1 megabit per second (Mb/s) throughput of I/O [33].

[28]MIPS is an acronym for Million Instructions Per Second.

[29]MFLOPS is an acronym for Million Floating Point Operations Per Second.

sponse time, the numbers are still interesting to computer engineers who want to capture a relative system performance and combine it with the system's speed to obtain the average number of instructions executed per machine clock cycle [44, 100]. MIPS can be expressed as:

$$MIPS = System\ SPEED_{MHz} * IPC_{ave} \qquad (2.2)$$
$$= \frac{1}{CT * CPI_{ave}} \qquad (2.3)$$

where $SPEED_{MHz}$ is the processor's speed in megaHertz and IPC_{ave} is the average number of instructions per clock cycle. Equation (2.2) shows that the MIPS metric is dependent on the machine's instruction set, making the comparison of MIPS ratings of machines with different instruction sets (CISC versus RISC) meaningless [33]. Furthermore, since the average number of instructions per cycle varies from one program to the other, the MIPS metric can vary on the same machine, making it a relative measure of performance and not an absolute.

2.2 Semiconductor ICs

Signal propagation is a major source of delay in a computer system. Smaller ICs and smaller boards result in smaller and faster machines, assuming that the functionality of the ICs stays constant. Functionality reflects the degree of integration in a chip and the complexity of the tasks it can perform. For example, current microprocessors perform integer, floating point, and memory management operations, making their functionality far superior to microprocessors ten years ago in which integer operations were the norms of IC integration. Decreasing the size of an IC implies reducing the die size. A die, sometimes referred to as chip, is a small unpackaged functional element made by subdividing a wafer of semiconductor material[30]. The wafer itself is laser-sliced out of a semiconductor ingot[31], the manufacturing of which must be very uniform and clean to reduce the number of wafer defects and, consequently, reduce the costs of the dies.

---

[30]The definition of the word "die" was obtained from the IEEE Standard Dictionary of Electrical and Electronic Terms (IEEE-SDEET).

[31]A semiconductor ingot is cylinder shaped, with a diameter equal to the wafer diameter; the companies that manufacture the ingots must pass certain high quality tests before they can be considered as ingot suppliers [72].

2.2. SEMICONDUCTOR ICS

This section discusses the technological and manufacturing trends for semiconductor ICs. Subsections 2.2.1 and 2.2.2 present the semiconductor physical characteristics and the *pn*-junction configuration, respectively. Subsection 2.2.3 describes the most utilized transistor fabrication technologies. Subsection 2.2.4 presents three technologies for increasing the IC speeds of operation while keeping the same die area. Subsection 2.2.5 provides two ways of improving the IC speeds of operation by decreasing the die area. Subsection 2.2.6 briefly discusses the issues of computer system reliability and fault tolerance.

2.2.1 Semiconductor Physical Characteristics

Semiconductors are electronic materials with a range of resistivity between insulators and metals: at high temperatures, they behave as conductors, and at low temperatures, they behave as insulators. Silicon is the most widely used semiconductor; it is cheap and readily available in nature as sand. Other semiconductors like germanium (Ge) and gallium arsenide[32] (GaAs) are used in the electronic industry, GaAs being an integral part of very fast switching circuits.

Silicon has a diamondlike crystal atomic structure [27]. It belongs to Group IV of the chemical periodic table; each atom has four valence electrons with other silicon atoms. The electron pairs of each crystal silicon atom are shared with other atoms with what is called a covalent bond. As the temperature of the silicon rises, electrons may be freed from the covalent bond and a hole created. A hole acts like a positive electron charge in a silicon crystal lattice where an electron is missing in one of the covalent bonds.

The process of hole and electron creation is referred to as ionization. An intrinsic semiconductor is a pure semiconductor, where the concentrations of holes and electrons are equal. An extrinsic semiconductor is a doped or impure semiconductor where holes or electrons are injected[33] in the silicon in order to create an imbalance in the charge concentrations. If the concentration of electrons is larger than the concentration of holes, the semiconductor is referred to as *n*-type or *n*-doped, and if the concentration of the holes is larger than the concentration of electrons,

---

[32]Gallium arsenide is not available in nature; it has to be formulated.

[33]Temperature increase can create a charge imbalance, but usually the semiconductor is operated at temperatures much lower than the ones that could create such an imbalance.

the semiconductor is referred to as *p*-type or *p*-doped.

2.2.2 *pn*-Junction

By joining a *p*-type and an *n*-type semiconductor or by diffusing *n*-type impurities into a *p*-type semiconductor or vice versa, a *pn*-junction is created. The *pn*-junction is the most important physical part and concept of the semiconductor electronic devices sector. It is at the core of the semiconductor transistor[34], the driving device of today's DRAMs, CPUs, and all the other application specific ICs (ASICs). For an historical background of the semiconductor industry, its economics, and the major players and their share of the world semiconductor markets, consult Yoffie [106].

2.2.3 Transistor Fabrication Technologies

A transistor is an active semiconductor analog device with three or more terminals [35]. Of the many transistor types utilized in ICs, the bipolar junction transistor (BJT) and the field-effect transistor (FET)—including CMOS[36], which is a FET-derived technology—are the most common. What follows is a description of the physical process of building these different transistor types.

Bipolar Junction Transistor

A bipolar junction transistor is formed by connecting two *pn*-junctions back to back. The configuration of a BJT can be either *pnp* or *npn* and, since the electrons have a higher mobility than the holes, the *npn* configured BJT is the most widely used in building ICs. Figure 2.2 presents the *npn* BJT configuration, where the *p* layer is referred to as the transistor's base and the outer and inner *n* layers are referred to as the transistor's collector and emitter, respectively. For more details on the BJT operation modes and circuit configurations, see Chapter 2 of Ghausi [27].

---

[34]The first solid state transistor was developed at Bell Telephone Laboratories in 1947 by William B. Shockley and his team [106].
[35]IEEE-SDEET
[36]CMOS is equivalent to Complementary Metal-Oxide Semiconductor FET.

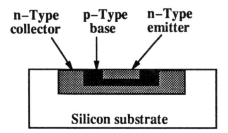

Figure 2.2: *npn* bipolar transistor configuration.

Field Effect Transistor

There are two basic types of FETs, junction and metal-oxide semiconductor FETs (JFETs and MOSFETs). The *pn*-junction and the electric field controlled current are the FETs' basic operational mechanisms. The MOSFET is the most widely used transistor in monolithic ICs[37] because it occupies less space[38], has a high input impedance (that is, draws less current and, eventually, consumes less power), has a low fabrication cost, and is less noisy than the BJT. There are two types of MOSFET, the enhancement and the depletion types, and each can have the main charge carriers as electrons or as holes. If the main carriers are holes, the MOSFET is called a *p*-channel MOSFET or PMOS, and if the main carriers are electrons, the MOSFET is called an *n*-channel MOSFET or NMOS. As mentioned earlier, electrons have a higher mobility than holes, so NMOS transistors are more frequently used as the building block of ICs.

Presented in Figure 2.3, an enhancement type NMOS transistor is built by diffusing two heavily doped *n*-type regions, referred to as source and drain, in a lightly doped *p*-type substrate. A silicon dioxide (SiO_2) layer covers the source and the drain, over which a metal plate, like aluminum or silicon, is deposited to form the FET's gate. To build a

---

[37]ICs are divided into two categories: monolithic ICs which are fabricated on a single semiconductor substrate and hybrid ICs where various components on separate chips are mounted on an insulating substrate and interconnected.

[38]In general, MOS transistors occupy the least space in IC fabrication and, on average, the ratio of MOSFET space to BJT space is 0.2, or smaller by a factor of 5.

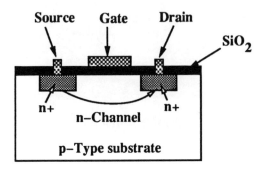

Figure 2.3: Enhancement type NMOS transistor configuration.

depletion type NMOS out of an enhancement type, a thin channel of lightly doped n-type material is diffused between the heavily doped n-type source and gate before covering them with the oxide layer. Of the two types, the enhancement is more widely used due to the simpler fabrication steps and the lower power consumption characteristics[39].

CMOS

Due to the low power consumption of enhancement MOSFETs and advancements in the IC manufacturing techniques, enhancement type PMOS and NMOS transistors can be fabricated on the same IC to obtain the complementary symmetry MOS or CMOS[40] inverter (see Figure 2.4). For more details on the operation modes and the CMOS circuit configurations, see Chapter 3 of Ghausi [27]. Due to its temperature stability, lower power consumption than NMOS, low cost of production, and high packing density, CMOS has captured a high percentage of the very large scale integration (VLSI) sector of microelectronics [49, 51, 75, 102]. However, one major drawback of CMOS is its limited switching speeds. The maximum projected speed for CMOS based ICs is in the 70 to 100

---

[39] A depletion type MOSFET consumes more power than an enhancement type due to the presence of the thin channel between the source and the drain. The channel in a depletion type MOSFET gives rise to leakage currents under static or direct current (DC) voltage conditions.

[40] The CMOS configuration was invented by Frank Wanlass in 1963.

2.2. SEMICONDUCTOR ICS

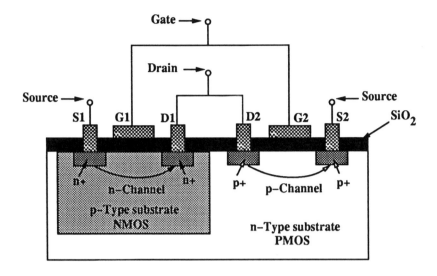

Figure 2.4: CMOS inverter configuration.

megaHertz range [64, 72].

2.2.4 High-Speed IC Technologies

For applications requiring higher circuit switching speeds, three alternatives exist: BiCMOS technology, GaAs technology, and optical technology.

BiCMOS

The BiCMOS fabrication technology is formed by combining the CMOS technology and the bipolar technology, known for its high switching speeds (100–200 megaHertz [64, 72]), high power consumption, and high fabrication costs. This alternative was driven by the bipolar technology manufacturers—who in the 1980s saw their market share slipping to CMOS—and by the proliferation of very fast cache memories (SRAMs) in microcomputers and workstations. New packaging techniques and materials and new cooling techniques are being developed to handle the high power consumption and the heat generated by BiCMOS ICs [49].

GaAs

This alternative uses gallium arsenide instead of silicon as the semiconductor. Gallium arsenide, formulated in the 1950s by Henry Welker of Siemens Laboratories, has an electron mobility of up to six times that of silicon, with circuit switching speeds exceeding the 1 gigaHertz (GHz) range; it uses less power than silicon and can convert electronic signals to light [10, 12, 24, 27]. Furthermore, GaAs ICs costs, which were once considered so outrageous that only the DoD[41] and supercomputer manufacturers could use them in their products, are decreasing significantly for three reasons [10, 12, 64]: (1) the GaAs manufacturers were able to map the silicon process technology to their GaAs ICs[42] manufacturing techniques; (2) larger GaAs wafers are being produced; and (3) higher production yields are attained in the manufacturing laboratories. Another attractive feature of GaAs is that it can become superconductive if cooled to a -263°C [24], which opens the door, previously barred by difficult manufacturing processes [37], to superconductivity and to superconductive ICs for integration in the supercomputers of the future[43].

Optical

The final alternative to increasing the ICs switching speeds is to use photonic, rather than electronic, transmission of signals. Photonic transmission involves optical interconnections and switches which can be obtained by laser diodes used as light transmitters and photo detectors [83]. Gallium arsenide, one of the prime materials in microwave circuits fabrication, can be used in the manufacture of the laser diode. Although the optical technology switching speeds can reach the gigaHertz range, it is still a relatively new technology and prohibitively expensive for use inside a commercial computer. Optical computers may be available by the end of this decade or the beginning of the next, but the transition will undoubtedly not be easy [83].

---

[41]DoD is an acronym for Department of Defense.

[42]GaAs was mostly used in satellite microwave applications, converting and amplifying microwave signals in the gigaHertz range to electronic signals.

[43]To keep supercomputers operating at normal temperatures, cooling processes are already being used to absorb the heat generated from their circuitry.

2.2.5 ICs: Speed Versus Die Area

After presenting an introduction to the semiconductor materials and technologies, it seems appropriate to discuss ways of increasing the speed of operation of a die by decreasing its area, while keeping the same functionality of the IC. There are basically two ways of accomplishing this, either by decreasing the feature size or by reconfiguring the IC layout.

Feature Size

As the feature size decreases, the ICs die areas decrease, making faster ICs switching speeds possible. The feature size is the minimum resolvable distance separating two etched lines in the semiconductor substrate. A line is etched in the semiconductor substrate by a process called lithography. Lithography is the process of transferring a circuit configuration on the semiconductor substrate. Of the many lithography methods, optical lithography is leading the electronics industry in the 1990s [21]. Optical lithography uses an ultraviolet beam to trace a pattern in photoresist. Photoresist is a substance that hardens when exposed to certain light frequencies—for example, ultraviolet. Developing the corresponding photoresist chemical structure for the particular wavelength of light has been a challenge to ICs manufacturers [21]. Nevertheless, when the wavelength of the light hitting the photoresist decreases, the feature size decreases. Once the circuit pattern is transfered to the photoresist, the non-hardened part is scraped off the semiconductor, and impurities are diffused in the silicon to create the doping characteristics described earlier. Details on the factors that influence the resolution in the photoresist and, consequently, the feature size of the etching process can be found in [21] and Appendix B of Ghausi [27].

Optical lithography[44] can reach a minimum feature size of 0.2 micrometer, and, considering that a gigabit (Gb) DRAM[45] chip requires a feature size of 0.15 micron, optical lithography may very well lead the semiconductor industry to a 256 megabit, if not a 512 megabit DRAM by the end of the decade. The other lithography contenders, like x-ray (with minimum feature size below 0.01 micron), electron beam, and ion

---

[44]Lithography or, indirectly, feature size, accounts for two-thirds of the increase in DRAM densities [21].

[45]DRAMs are the test bed of any increase in feature size because they are the easiest chips to manufacture.

beam (both with minimum feature sizes below 0.1 micron [21]), do not yet have the process maturity and high yields to be used in the production lines. Large companies such as IBM and AT&T are pushing these technologies, hoping for a big payoff by the turn of the century [21].

Circuit Layout

Another way of decreasing the die area is by reconfiguring the layout of the IC and optimizing the number of connections among the different subcircuits embedded in the chip. But the layout problem and the min-cut problem are NP-complete[46] [25, 48], making this option even more difficult to tackle than decreasing the feature size. (Min-cut refers to the minimum number of connecting lines cut in partitioning an IC to reconfigure its layout.)

2.2.6 Reliability of Computer Systems

Feature sizes in the submicron range are being etched in 1+ square centimeter dies. Several of these dies are assembled together to form the processing board, the memory board, the hard disk controller, the display image driver, and the printer driver. All of the previous assemblies require high fault tolerance and reliability. If a computer system breaks down, the problem could be at the assembly level or at the ICs manufacturing submicron level. Designing fault tolerant and reliable ICs is a major goal not only because of the costs involved in repairing a faulty IC, but in response to the sensitivity of consumers who are buying a machine the operation of which cannot be seen by the naked eye.

2.3 Magnetic Hard Disks

Storing digital information safely and retrieving it efficiently have been two of the most important attributes of a computer system's storage device. Regardless of how fast the machine processes data, the time the processor spends waiting on data and the accuracy of the information retrieved will always be the most important factors affecting the throughput of the machine and the validity of its outputs.

---

[46]NP is an acronym for Non-Deterministic Polynomial. It has been conjectured that all NP-complete problems are intractable [25].

2.3. MAGNETIC HARD DISKS

This section elaborates upon magnetic hard disk storage technologies. Subsection 2.3.1 defines a storage system within a computer and a computing environment. Subsection 2.3.2 describes the magnetic storage technology with a brief presentation of its history. Subsection 2.3.3 presents a description of the magnetic hard disk components, their functions and physical structure. Subsection 2.3.4 provides a trace of a hard disk data request and an analysis of the factors affecting its performance, in particular the hard disk's response time. Subsection 2.3.5 provides some concluding remarks about the magnetic storage technology and about the importance of VLSI as one of the main drivers of the storage technologies.

2.3.1 Computer Storage System

A storage system is considered external to the communication between the processing board and the main memory board, and nonvolatile because it retains the stored data even when the power to the system is switched off. The most widely used nonvolatile storage technologies are magnetic and optical based. With today's state-of-the-art technology, however, magnetic storage has the edge with respect to price, response time, and storage density [103]. It is the storage technology discussed most fully in the following subsections.

The storage system is accessed by an operating system instruction as a result of a user directory read or write command, a process load instruction by the user, or a data-access miss requested by the processor from the main memory during the execution of a certain program. Whether the storage system is a single user or a shared system, each user has a specific storage space within a specified directory of the system. The directory is allocated by the operating system on request by the system's manager or the user[47]. The user can write his/her working files to the allocated directory, read these files, or load programs from it into main memory and execute them.

2.3.2 Magnetic Storage Technology

A magnet is a polarized ferromagnetic material. The polarization has the direction of the magnetizing field, either a north–south (N–S) direction[48]

---

[47]In certain cases, a directory is allocated on request by a running process.

[48]The north–south direction refers to the orientation of the magnet's poles.

or south–north (S–N). A permanent magnet retains the polarization even after the magnetizing field is removed and this accounts for the nonvolatility of a magnetic storage system. The digital data bits are stored as polarized magnetic particles, with a 0 bit equivalent to a S–N polarization and a 1 bit equivalent to a N–S one, or vice versa. There are several types of magnetic storage systems, such as magnetic tapes, hard or rigid disk drives, or floppy diskettes. Of the three, the hard disk drives possess the highest data-access rates and best complement a workstation's high-speed processor. Hard disk drives also have high linear and areal densities, useful attributes which will be described later.

Magnetic Storage History

IBM[49] pioneered the direct access storage devices (DASDs), in particular the magnetic hard disk technology that we know today [58]. The areal density of DASDs increased by a factor of 100 in the last 20 years [58]. New developments and innovations continue to drive the price per bit of storage down at a rate of 15 percent to 20 percent, compounded yearly [3]. Data reliability, performance, and low cost per bit continue to be the main attributes of this technology. Of these three attributes, performance is the most important because it reflects the speed of access to the data stored in the system and dictates the mechanical and media configurations used in the storage device.

Before discussing the factors influencing its performance, a description of the magnetic hard disk basic components setup and operational procedures follows [58, 61, 80].

2.3.3 Magnetic Hard Disk Components

Figure 2.5[50] shows the enclosure of the storage system containing a stack of rigid disks mounted on a spindle, their corresponding read/write/erase heads with their sliders, and the electromagnetic actuator which controls the movement of the heads over the disks' surfaces. Not shown in the figure is the data channel that feeds data from main memory to disk and vice versa. The air inside the enclosure is filtered so that no particles can interfere between the disk media and the heads during the read/write

---

[49]IBM is an acronym for the International Business Machines corporation.
[50]C.D. Mee and E.D. Daniel, **Magnetic Recording Handbook**, ©1989 McGraw-Hill. Reprinted with permission of McGraw-Hill Book Company.

2.3. MAGNETIC HARD DISKS

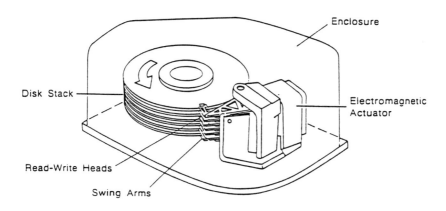

Figure 2.5: Rigid disk file components. Source: [58]. ©1989 McGraw-Hill.

operations. To control the heat generated during the disks' rotation, an air cooling mechanism is used [58]. The disk substrates are mostly made out of aluminum-magnesium alloys. Glass or ceramic substrates are used for extremely fine head to medium spacings because they provide a smoother surface on which to deposit the magnetic media [58, 80]. In addition, a thin electroless layer of nickel-phosphorus is plated on the substrates and polished before depositing the media to make the disks surfaces even smoother and flatter. Since the disks rotate at a specified constant speed, their thickness is chosen within the resonance requirements of the disk assembly [58]. Any mild resonance or vibration can induce reading errors and write misregistrations in the system. The coating on the disks[51] is an alloy of some highly coercive[52] ferromagnetic iron oxide, the layout of which is recognizable by the read/write/erase mechanism of the device. The magnetic layers are deposited on the disks in the form of small, needle-shaped particles or as a magnetic film[53], and the data stored on them will be retained as long as the disks are not exposed to high heat or fluctuating magnetic fields. To shield the media from air particles and from the heat generated by the disks' rotations, a thin layer of hard carbon overcoat is deposited on top of the media-coated disk surfaces [103].

Magnetic Disk Surface Layout

As indicated in Figure 2.6, each magnetic oxide medium is laid in concentric tracks, separated from one another by a constant pitch. An emerging technology uses IC process technologies to etch the tracks on a disk's surface[58] and, consequently, makes possible both finer track pitches and more clearly defined tracks. (A nonhomogeneous coating layout could induce reading and writing errors due to the noise generated from the proximity of the bit cells and the tracks [58].) Each track is divided into an equal number of sectors, and each sector[54] stores the same number of data bytes[55] regardless of the sector's radial length[56].

---

[51] The disks can be coated on both sides.
[52] Coercivity is the magnetic field required to reduce the magnetization of a bit to zero.
[53] Most magnetic films are cobalt-based metal alloys.
[54] The data on each track are updated one sector at a time.
[55] A byte of magnetic data is a series of eight magnetic cells.
[56] In some disk storage systems where the data channel's bandwidth is not constant, the data recorded on each sector is proportional to the sector's length [33].

2.3. MAGNETIC HARD DISKS

Since the disks are rotating at a constant speed with one head per disk surface, and since the data channels usually have a constant bandwidth, the data access rate is constant and the number of bits per track is constant. The track density, then, is limited to the density of the innermost track. The innermost track's radius depends on the radius of the spindle support and rotation mechanisms. For a constant data rate, maximum disk areal density is obtained for an innermost track radius equal to half the disk's radius (see section 2.2 of Mee and Daniel [58] for a detailed proof).

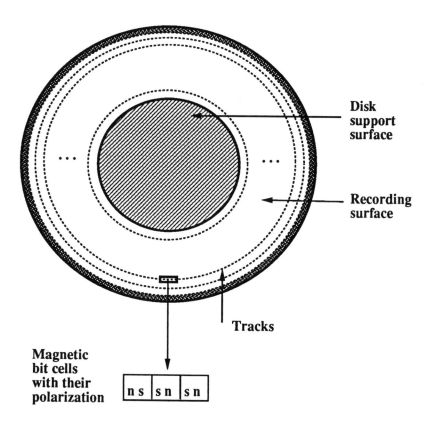

Figure 2.6: Magnetic disk surface layout.

Read/Write Heads

The magnetic heads are attached to slider arms, each dedicated to a disk surface, and supported within a submicron distance from the rotating media by a hydrodynamic air bearing [58]. Most heads are inductive electromagnets and, as illustrated in Figure 2.7, an inductive head is shaped as a ring with a small gap at the surface facing the magnetic medium. The core of the electromagnet is either ferrite or metallic film based. A coil is wrapped around the core, the functionality of which will be described shortly.

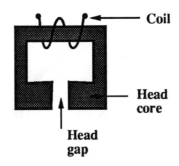

Figure 2.7: Inductive head structure.

Heads having the previous configuration can be used for reading, erasing, and writing data to the media [58]. Ferrite cores are easy to manufacture and have improved wear characteristics. However, at high frequency operations, an all-ferrite core prevents a magnetic flux from penetrating it, a problem which some have suggested might be eliminated by laminating it with an insulator like alumina [103]. The resulting head is referred to as the metal-in-gap (M-I-G) ferrite head.

Thin film metal (TFM) based heads have the same configuration as the ferrites, but they are not as easy to manufacture. A photolithography process is used to deposit thin layers of metal on a thick ceramic wafer; at the end of the process, the wafer is cut into small dies which are packaged later as TFM heads [3, 5, 90, 103].

Regardless of the materials used in its construction, the inductive head operates in the same basic way to read from or write to the disk.

What follows is a description of the writing and reading processes of inductive heads [58]:

The Writing Process. Since the heads are on a spindle, they can move all together to the corresponding tracks, and a sector[57] can be written or read on each disk surface simultaneously; but for the purpose of illustration, the writing process of one head on one disk surface will be described. As the disks rotate at a constant speed and the head reaches the corresponding sector on the specified track, a temporary change in the coil[58] current induces a magnetic field in the head's core. The direction of the magnetic field in the core depends on the direction of the current in the coil; one current direction corresponds to a zero bit being registered and the other a one. With a specified direction, the magnetic field attempts to cross the core's gap, polarizes the magnetic bit cell passing under the head, and a bit is written[59].

The Reading Process. As the disks rotate at a constant speed and the head reaches the corresponding sector of the specified track, a magnetized bit cell, passing under the head's gap, induces a magnetic flux in the core. In turn, the changing core flux induces a certain voltage in the coil. The value of the voltage is interpreted by the disk's circuitry and is dependent on the magnetic bit cell read. For a more detailed look at different read and write mapping techniques on magnetic hard disk media, consult section 2.2 of Mee and Daniel [58].

Both, the M-I-G and the TFM heads are suitable for high data rates and high bit density systems, and the fact that they are not touching the rotating disks adds to the reliability of the stored data and to the durability of the media. The number of heads per storage system depends on how many disks the system has, whether each side of the disks is magnetically coated, and whether each head can perform read, erase, and write functions. A head per track configuration was suggested at one time in the past, but dropped later because its implementation was not

---

[57] As mentioned earlier, writing and reading from a rigid disk is performed one sector at a time and not one byte or several bytes at a time; a sector is a multitude of bytes, the number of which is dependent on the storage system.

[58] The coil is wrapped around the head's core.

[59] The erasure of the bit is overridden by the writing process.

cost competitive with the DRAMs [58]. For extremely high bit densities and high data rates disk systems, separate read and erase/write heads were suggested, each with its optimized functional capabilities. Magnetoresistive read-only and inductive write-only heads have been developed for these purposes and are currently implemented in IBM's top-of-the-line hard disk storage systems [103]. For more details on the operations of the dedicated heads, refer to [3, 5, 58, 90, 103].

Actuator and Servomechanism

The heads' movement over the rotating disks is a mechanical action, controlled by a head-positioning servomechanism. The servo determines the distance the heads must travel until they reach the right tracks, initiates the heads' movement, and stops the heads when the specified tracks are reached. In most rigid disk systems the location of the data on the disks is recorded on one disk, called the servo disk [58]. When a read or write command reaches the servo, the location of the data to be read or written is determined from the servo disk, and an electromagnetic actuator is instructed to move the corresponding disk heads to the specified track coordinates. Also, the actuator must keep the heads on track to minimize misregistration or misreading errors [58]. A discussion of the actuator's design and the servomechanism's circuitry is found in section 2.2 of Mee and Daniel [58].

2.3.4 Magnetic Hard Disk Performance

The performance of a magnetic hard disk is a measure of how fast it comes back with the requested data. The response time of the hard disk has embedded in it all the attributes of the mechanical, electrical, and magnetic properties of the storage system. What follows is a trace through a data request and an evaluation of the factors that affect the storage system's response time.

Stepping Through a Data Request

The hard disk has a FIFO[60] queue for incoming data requests. The top request in the queue is accessed by the disk's controller, which translates it into a servomechanism command. The servomechanism finds the

---

[60]FIFO is equivalent to the First-In First-Out paradigm.

2.3. MAGNETIC HARD DISKS

corresponding sector by searching for it in the servo disk[61]. Once the corresponding disk and the sector location on it are found, the servomechanism computes the corresponding distance the disk surface's head has to travel from its current position. The travel distance of the head is then passed to the electromagnetic actuator, which controls the head movement until it reaches the corresponding track[62]. Once the target track is reached, the head has to wait for the corresponding data sector to come under it[63]; the head may miss the sector. If so, it might have to sit and wait one disk revolution before it reaches the specified sector a second time. Once the sector is found, the data transfer occurs, either a data read or a data write[64].

Magnetic Hard Disk Response Time

After tracing a hard disk request procedure, the response time of a magnetic hard disk storage system is expressed as follows:

$$T_{response} = T_{service} + T_{wait} \quad (sec) \qquad (2.4)$$

where the wait time is the time the request had to wait in the queue. The service time's expression is:

$$T_{service} = T_{actuator} + T_{seek} + T_{latency} + T_{miss} + T_{data-transfer} \quad (sec). \qquad (2.5)$$

The service time in equation (2.5) has been improved over the years by innovations and enhancements to the mechanical, electrical, and magnetic components of the storage system, for instance:

1. The data transfer and the actuator positioning times have been improved by using a harmonious combination of mechanics and electronics, called micromechatronics, to put together the head, the rotational motor, and the control servomechanism into one single device [90].

2. The magnetic bit cell length has been decreased and with it the gap width of the head's core decreased, the linear densities of the media[65] increased, and the data transfer time decreased. A new

---

[61] The time to termination of this procedure is referred to as the actuator positioning time.
[62] The time to termination of this procedure is referred to as seek time.
[63] The time to termination of this procedure is referred to as latency.
[64] The time to termination of this procedure is referred to as the data transfer time.
[65] Linear density is expressed as the number of bits per millimeter.

bit cell layout technique aligns the cells perpendicularly instead of longitudinally along the track, which increases the bit cell densities. (This layout technique was not possible in the past due to unavailability of media that could handle such a magnetization [58].) Unfortunately, the increase in the bit cell densities caused bit interference noise and bit misregistration and misread problems during hard disks accesses. To preserve the storage system's data reliability, smaller head designs with very small core gaps were implemented [50, 58, 103]. The noise problem was handled by decreasing the head medium spacing [58, 103].

3. The track density[66] has been increased by improving the head technology and the layout process of the tracks. Magnetic film media and ICs etching processes have been used to obtain a finer track layout and a smaller track pitch [50, 58]. The increase in track densities induced an increase in the areal density of the disk—equal to the product of the linear bit density and the track density—and a decrease in the seek time. Unfortunately again, the increase in track densities caused track-to-track noise interferences during a write or a read operation, problems that were solved by better actuator and servomechanism designs and by the development of smaller heads with very small core gaps [50, 58, 103].

4. Multiple track access implementations, coupled with faster disks rotational speeds, have steadily decreased the data transfer time [50, 58]. With the increase of the data transfer rate[67], the channel circuitry has improved and faster data processing ICs have been employed to handle the higher bandwidth between the storage system and the main memory.

2.3.5 Concluding Remarks

Until very recently, paper was the dominant form of storage [50]. The digital information storage technologies have, nonetheless, become cheaper, more reliable, more easily accessible, and more user friendly, encouraging business, government, and private sectors to rely upon them more and more.

---

[66]The track density is expressed as the number of tracks per millimeter.

[67]The data transfer rate is expressed as the number of megabytes per second.

2.4. COLOR CRT DISPLAYS

Note once again, however, that VLSI technologies are integral parts in the head manufacturing process and the disk control circuitry. Lithography is being used in the manufacture of thin film heads and digital signal processors, and caches are being used in disk controllers to reduce the number of wait-on-data cycles of CPUs [107]. Since DRAM speeds have not kept pace with microprocessors, further improvements to the storage attributes might help to compensate for lag in DRAM performance improvements, allowing the overall system performance to keep increasing.

2.4 Color CRT Displays

With the introduction of multitasking[68] capabilities into the workstation environment, the need for larger, faster, and appealing computer terminals emerged. The display technology best positioned to take the lead in the merger of computer and display hardware was cathode-ray tube (CRT) based. What follows is a brief presentation of the CRT's history (subsection 2.4.1), a description of other display competing technologies (subsection 2.4.2), a presentation of the color CRT components (subsection 2.4.3), an illustration of the generation techniques of a screen image and its elements (subsection 2.4.4), a presentation of the factors affecting the CRT's bandwidth (subsection 2.4.5), a description of the CRT's graphics adapter components (subsection 2.4.6), and, finally, a presentation of the CRT's resolution and the technical barriers facing the CRT industry as it attempts to achieve maximum resolution (subsection 2.4.7).

2.4.1 CRT History

The CRT's invention in 1879 is credited to William Crookes [92]. More than a century after its invention, CRT technology still has the highest market share in the display industry [93], receives the highest research and development (R&D) expenditures, and demonstrates the greatest improvements per R&D dollar spent [23]. CRT-based displays replaced the typewriter and the teletype terminal displays of the 1960s because

---

[68]In multitasking, a single user can interact with several computer applications concurrently.

they facilitated user interaction with time-shared computers[69], and had the fastest response times [20]. Several computer-aided design (CAD), animation, simulation, and layout applications were developed in the late 1970s, and only the CRT display provided the color imagery necessary to capture the illusion of reality in animations, simulations, and designs. Its large screen size enhanced the multitasking attributes of the machine[70] and delivered the fast response time that provided for an efficient interaction between the user and the computer [56].

CRTs are available in black and white (B&W) and in color image display capabilities. The first color CRT was produced in 1950 [92], and had the B&W CRT technology as its backbone. Today, the CRT still holds the highest market share among the total display technology market—71 percent as of 1989—and it is not projected that other technologies will challenge its reign until the mid-1990s [93].

2.4.2 Competing Display Technologies

Several display technologies competed with the CRT technology for market share in the past, but it was not until the age of the personal computer that the market witnessed a flood of innovations from display companies around the world. Of the display technologies competing with the CRT's, a few show some promise, and Japan is the leader in their development and market share [23]. The most promising of these technologies is liquid crystal (LC) based. LC displays (LCDs), available in black and white and in color, are compact and low in power consumption, are increasing in size and decreasing in cost, and their display elements are easily addressable with thin film transistors. But improvements are needed to increase their resolution and to enhance the speed of their picture updates for animation [23, 45, 79]. The other two potential technologies are electroluminescent displays (if full color ones are developed) [88] and plasma displays (once their technology matures and affordable color capabilities comparable to the CRT's are achieved) [23]. For a more detailed discussion of these competing technologies, see [23, 45, 79, 88, 92, 93].

---

[69]Time-sharing is an operating system function that schedules the processing time of the computer among several users, each with a particular computer time slot.

[70]Controlled by a window manager, several user applications can be launched, each in its separate window, and with all the windows showing on the large screen.

2.4. COLOR CRT DISPLAYS 39

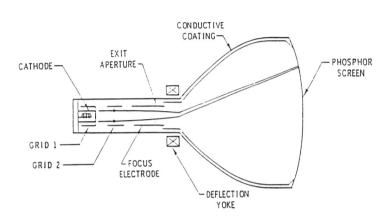

Figure 2.8: CRT display components. Source: [92]. ©1985 VNR.

2.4.3 Color CRT Components

As shown in Figure 2.8[71], a CRT consists of several parts [76, 92]: the bulb's[72] faceplate or viewing screen upon which phosphor dots are deposited, the funnel, and the neck. In addition, in color CRTs, a metal shadow mask is suspended behind the faceplate, between the electron beams and the color phosphor dots. The metal shadow mask, not shown in Figure 2.8, acts as the coordinate system for the image displayed on the screen of the CRT. What follows is a description of the previously mentioned components and their role in producing the image displayed on the CRT's screen.

---

[71]Lawrence E. Tannas, Jr., **Flat Panel Displays and CRTs**, ©1985 VNR. Reprinted with permission of Von Nostrand Reinhold Company, Inc.

[72]The bulb is also referred to as the bottle or the envelope.

Bulb's Faceplate

The bulb's faceplate defines the surface of the viewing screen and isolates the inner CRT components by maintaining the vacuum needed for the cathode rays to hit the inner side of the screen without any particle interferences. Normally, the vacuum-filled bulb is made of glass. However, in certain cases, the funnel and/or the neck may be made of ceramic or metal, depending on the CRT application [92]. In most cases, the screen is designed such that a 4:3 aspect ratio is satisfied, where the horizontal side of the screen measures 4/3 times the vertical side. The size of the screen, usually measured as the screen diagonal, is limited by the ability of the vacuum filled glass to withstand the atmospheric pressure on its faceplate. The vacuum in the bulb is maintained at a specified level by an antenna getter[73] [55].

Bulb's Neck and Funnel

The second major part of a color CRT is the neck of the bulb. The neck area houses three cathode heaters, three cathodes, three electrostatic beam[74] accelerating apertures, and an electrostatic beam focusing aperture. In certain cases, the beam focusing aperture is magnetic and positioned outside the tube's neck. Three electron beams are needed to excite three types of phosphors—red, green, and blue, simultaneously—and create the corresponding color on the viewing screen by an additive effect [92], a process that will be described later.

A cathode ray is generated in the vacuum by heating a barium compound, or a barium oxide *cathode*[75], to temperatures that enable the release of a high density *electron beam*, and by applying an electric field to the hot surface of that cathode. The electric field is generated by applying a positive voltage or potential to the anode of the *accelerating aperture*. The accelerating aperture has two parts: the first is the accelerating anode, which generates the positive potential to draw the electrons from the cathode surface, and the second part is the negative potential electrode, which controls the intensity of the electron beam. The negative potential electrode is sandwiched between the cathode and the accelerator anode and controls the beam intensity by applying a

---

[73] A getter is defined in SDEET as a substance introduced in an electron tube to increase the degree of vacuum by chemical or physical action on the residual gases.

[74] The beam referred to is the ray of electrons generated by the electrode.

[75] A cathode is an electrode at which negative charges are formed.

2.4. COLOR CRT DISPLAYS

negative potential electric field on the surface of the hot cathode; that is, as the negative potential increases, three consequences ensue: first, the repelling forces of the negative field on the cathode increase; second, the anode's positive field effect becomes weaker; and third, the intensity of the electron beam decreases. (Chapter 6 of Tannas [92] contains a discussion of several other accelerator structures.) The intensity of the beam affects the luminance and the resolution of the CRT, effects which will be described later.

While still in the bulb's neck, the electron beam leaves the acceleration aperture and goes through a *focus aperture*. The focus aperture can be magnetic[76] or electrostatic, but its task is to orient the beam toward the screen by overcoming the electrons scattering[77] and drifting[78] effects. Several focusing paradigms are discussed in detail in Chapter 6 of Tannas [92]. While leaving the bulb's neck, the electron beam passes through a *deflection aperture*. The deflection can be induced magnetically or electrostatically, and it is applied as vertical and horizontal fields on the beam to position it to the desired coordinates on the viewing screen. The viewing screen is maintained at a high positive potential and acts like an anode for the drifting electrons through the *funnel* of the bulb, which constitutes the third CRT component.

Screen Phosphors and Electron Beams

The geometrical positioning of the three electron beams with their corresponding generators, accelerators, and focus apertures depends on the color phosphor dots layout on the screen. These phosphors constitute the fourth major component of the color CRT. The array of colors on the screen is generated by adding three color sources: red, green, and blue (RGB). The phosphor RGB color sources pass through two stages to establish the cathodoluminescence phenomenon:

First Stage. As soon as the deflected electron beams drift into the funnel space (shielded from the earth's magnetic field by an internal canceling magnetic field) and hit the corresponding phosphor coordinates on the screen, the phosphors are excited by the energy with

---

[76] Magnetic focus apertures have had the highest resolution performance results.

[77] Scattering occurs due to the electrons' or the similar charges' repulsion effect.

[78] Drifting occurs due to the positive potential field while passing through the accelerator anode.

which the electrons are bombarding them and emit a fluorescent radiation.

Second Stage. Soon after the excitation ceases, the phosphors emit a phosphorescent radiation, completing the cathodoluminescence phenomenon.

Each of the three electron beams is dedicated to phosphors of a particular color, and all the phosphors on the screen have to go through that excitation process for an image to be created. The colors generated depend on the beams' intensity which can be controlled by the accelerating apertures of the CRT. As the beams drift in the funnel space, though, they experience scattering effects due to the electron repulsion phenomenon, which brings about the need for a fixed and defined phosphor coordinates scheme.

Metal Shadow Mask

The screen coordinates system takes the form of a metal shadow mask perforated with holes or slots, as shown in Figure 2.9[79]. This mask is placed right before the phosphor-covered screen and constitutes the fifth major CRT component. The perforated shadow mask works as follows[80]. Each hole in the mask covers an RGB triad of phosphors. The triads in front of the mask are arranged uniformly by a photoresist process. The photoresist process enables the layering of phosphor dots, one color at a time, where the photoresist covers the previous layer or layers of dots while one of the remaining layers is deposited. The latest phosphor-depositing-process technology uses thin-film phosphors because of their homogeneous surface and the image enhancements they provide to the CRT [81, 92]. Thin-film phosphors are mostly inorganic crystalline materials, with a few added impurities to enhance their activation efficiency during the electron excitation process. (Refer to Chapter 6 of Tannas for details on the different types of phosphors and their attributes [92].) After applying the phosphor film on an optically smooth faceplate, a thin

---

[79] E.W. Herold, "History and Development of the Color Picture Tube," **Proceedings SID 15(4)**, ©1974 SID. Reprinted with permission of Society of Information Display.

[80] A slotted mask configuration has been used in the Trinitron CRT technology developed by Sony Corporation. It is not described here.

2.4. COLOR CRT DISPLAYS   43

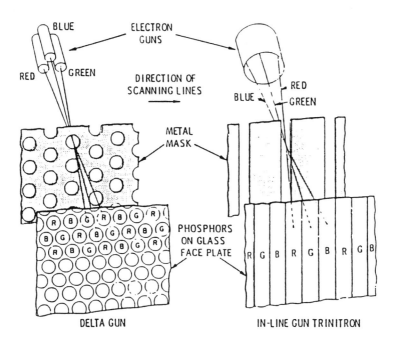

Figure 2.9: Inner screen phosphors layout structures. Source: [35]. ©1974 SID.

coating of aluminum is deposited on the film to enhance its conduction [92].

Figure 2.9 shows the electron guns positioned in a delta configuration to excite simultaneously the triangularly shaped RGB phosphor triads on the screen. As soon as the beams hit the phosphors, each with a particular intensity, the additive light emitted from the phosphors forms the final transmitted color of the image's dot. The limit on the number of phosphor dots that can be excited on the screen is a function of the metal mask's hole pitch. The smaller the hole pitch, the harder it is to manufacture the mask[81] and the more expensive the CRT becomes. By providing the mask, the definition of the image displayed is a function of both how well the circuitry of the CRT can focus the beam to hit the screen dots through the mask holes and the total number of holes drilled in the mask.

2.4.4 Screen Image

The beams can scan all the holes on the screen in two ways. The following paragraphs illustrate the two image generation techniques and describes the basic image element, the pixel.

Image Generation Techniques

The first image-generating technique is called interlaced scanning, where every other horizontal line of holes is scanned in the first pass of the beams and the nonscanned lines are scanned in the second pass. A line is equivalent to the number of holes perforated in the mask horizontally. Noninterlaced scanning is the second image generating technique used, and, in this scheme, no lines are skipped while scanning the screen. The electron beams scan the screen in one fixed direction; after scanning a line, the electron beams go into a blanking stage before they start scanning the next line. Due to flicker and phosphor light emission persistency limitations, each image on the screen has to be scanned, or refreshed, several times a second, depending on the CRT cathodoluminescence parameters. When interlacing is used, an image is scanned half as many times as the frequency of each pass (that is, if each pass is performed 80 times, with a two-pass full image scan, the whole screen is effectively refreshed 40 times). Most CRTs today use noninterlaced scanning because

---

[81]Smaller hole pitches induce lower yields of good masks.

2.4. COLOR CRT DISPLAYS

it provides sharper images and the available screen-driving hardware is capable of performing it, and it is cheap enough to be incorporated as a basic component of the workstation system [76, 92].

The Pixel

Each image is composed of picture elements, referred to as pixels. It is important to note that the image is in the computer memory and is being fed to the screen by the CRT's hardware driving circuitry. Each pixel is mapped onto the computer memory as a series of bits. The bits are transformed into analog signals that control the intensity of the electron beams and, in turn, the color of the pixel.

2.4.5 CRT Bandwidth

The rate at which the pixels' bits are fed to the controls of the electron guns determines the rate by which the screen can be scanned, or the bandwidth. The bandwidth is equivalent to the rate at which the electron guns can be switched on and off. Special attention must be given to high switching electron guns rates due to the overheating that occurs in the deflection apertures [60]. Several cooling techniques have been employed in the more sophisticated CRTs [60, 92]. The guns are assumed to switch on and off every time a pixel is scanned because different beam intensities are supposed to be generated.

The maximum bandwidth is equal to the number of holes in the shadow mask multiplied times the refresh rate of the screen. The refresh rate of the screen is the number of times the screen is scanned per second to avoid flicker. The maximum bandwidth is expressed in equation (2.6) as:

$$BW_{max} = (HS_{in} * HPI) * (VS_{in} * HPI) * RR_{Hz} \; (Hz) \quad (2.6)$$

where HS and VS are the horizontal size and vertical size in inches of the CRT screen, HPI is the number of mask holes per inch, and RR is the refresh rate in Hertz. The number of holes in the metal shadow mask, or the maximum number of pixels that can be displayed, is equal to the number of holes in a horizontal line multipled by the number of holes in a vertical line.

2.4.6 CRT Hardware Drivers

Hardware drivers of a color CRT are components of the graphic subsystem of the computer workstation, and they are typically assembled on a circuit board placed in the workstation cabinet or in the CRT assembly itself [9, 68, 104]. They are made up of four components. The first component of the hardware drivers is the *controller*, which handles the interface between the computer and the peripherals, such as the keyboard or the mouse, and computes the coordinates of the pixels to be echoed later as the scanned image on the screen. The second component is the pixel memory storage, or *buffer*, which is constantly updated and accessed in the refresh cycles of the screen. The availability of very fast DRAMs allowed CRTs to replace the display technologies of the 1960s like teletypes and typewriter terminal displays [56]. The third component is the *video generation system*, responsible for feeding the pixels' serial bit stream to the D/A[82] converter to generate the RGB color analog signals, for refreshing the screen, and for synchronizing the CRT's response to the analog signals just created, in particular the electron guns' intensities. The fourth component is a *supervisory processor* which coordinates the timing and the execution of the tasks previously mentioned. Several hardware drivers can be designed for specific applications of the CRT, and these drivers could become a major cost overhead to the total cost of the workstation [33]. A recent enhancement of the hardware driver's attributes has been achieved through the use of cache memory to accelerate the image refresh rates and timing and to improve the resolution of the display as a whole.

2.4.7 CRT Resolution

Resolution has become the key parameter in any display's technical description, in particular CRT-based displays. Estimated to double every five years in the future [77], resolution captures the performance characteristics of all the CRT's major components and the hardware driving it [98]. The Standards and Definitions Committee of the Society for Information Display defines resolution as a measure of the display's "ability to delineate picture detail." In other words, it is a measure of the number of pixels displayed per image [92].

---

[82] A D/A is a digital to analog converter.

2.4. COLOR CRT DISPLAYS

To increase the CRT's resolution, several factors and technical barriers must be taken into consideration including: (1) the hole pitch of the metal shadow mask, (2) the switching speeds of the electron guns, (3) the expansion of the metal shadow mask, and (4) the misconvergence of the electron beams.

Metal Shadow Mask Hole Pitch

The maximum attainable resolution is equal to the number of perforated holes in the metal shadow mask. So, for a fixed screen size CRT, increasing the maximal resolution is synonymous with decreasing the mask's hole pitch. But the smaller the hole pitch, the harder it becomes to manufacture the mask. The maximum resolution per inch that a human eye can discern is equal to 300 holes per inch [41], or a hole pitch[83] of 0.085 millimeter. For a color CRT with a 19-inch diagonal and a 4:3 aspect ratio the maximum resolution requires for nearly 15.6 million holes to be drilled in the mask, or approximately 4,000 holes on the side for a 1:1 aspect ratio! By today's standards, drilling such a huge number of tiny holes in the metal shadow mask is extremely difficult. The yield of good and nondefective masks at such a high resolution is so low that the manufacturing costs of the corresponding color CRTs are prohibitively high.

Higher resolutions might be economically feasible for very small color CRTs [40], or B&W CRTs where no shadow mask is required [77]. But for 14- or 19-inch workstation displays, several technical problems other than mask yield remain.

Electron Guns Switching Speeds

The second technical barrier to increased CRT resolution is switching the electron guns quickly enough so that no flicker occurs at such high resolutions [41, 77]. Given the 19-inch color CRT screen with 300 holes per inch resolution, the maximum bandwidth of the guns is close to 1.56 gigaHertz for a screen refresh rate of 100 Hertz. Today's multi-guns structure and design are not adequate to handle these switching speeds, and several material reformulations are needed to handle the amount of heat generated from the screen scanning process [13, 41]. Primarily, the phosphors that are used need enhancements. At such a high bandwidth,

---

[83]The shadow mask hole pitch as of 1991 is 0.28 millimeter.

the phosphor dots are being excited for a very short period of time, so only phosphors with smaller electric current[84] requirements will satisfy the maximum resolution limitations.

Metal Shadow Mask Expansion

Developing metal shadow masks that will not expand as their temperature increases is also problematic in increasing CRT resolution. As the electrons bombard the screen through the mask, heat is generated and induces metal expansion [13].

Electron Beams Misconvergence

The fourth obstacle to maximizing resolution for workstation size displays is the electron beams' misconvergence [66]. Since the phosphor dots are smaller, the separation of the electron beams has to be increased to focus properly on the dots. However, this larger separation of the guns creates beam misconvergences on the screen.

Depending on the CRT's application, tradeoffs have to be considered in the final design to determine the degree of focus sacrificed for greater convergence of the beams. By today's standard designs, cathode materials will not pose a problem in attaining such high switching speeds [77]. The hardware drivers will not be an obstacle either, because VLSI's evolution brings with it faster microcontrollers, faster and cheaper DRAMs, and more efficient paradigms to refresh the screen at high rates for a large number of pixels [8, 14].

2.4.8 Concluding Remarks

The CRT belongs to the class of active, direct-view refresh displays. Its performance and cost are affected by the type of metal and hole pitch of the shadow mask, by the screen phosphors' brightness[85], persistency, and deposition technique, by the hardware driving circuitry and how well it matches the resolution and bandwidth parameters of the CRT, and by the overall design of the bulb [13].

The disadvantages of a color CRT are its high power consumption, its curved face plate, its volume, and its weight. Over the years, its high

---

[84]The electric current is equivalent to the electron beam.

[85]Brightness describes the perpetual effect generated by the luminance and the chrominance of the light emitting object.

switching speed, its high resolution, and its low cost have kept its market share high. It will remain the technology of choice for many workstation applications, especially in CAD and simulation markets. Currently, Japan is the leader in enhancing the CRT attributes [23] and, from the R&D money already invested in it, the CRT will most likely retain a high market share through the 1990s [16].

2.5 UNIX Operating System

The 1990s wave of information processing is fueled not only by advances in hardware capabilities—in the form of faster processing speeds, faster and denser memories, and high bandwidth communication links—but also by open system oriented software. The ability to develop open system software has been the key to creating new waves in the industry because it promotes the notion of standardizing the interface between different computer systems and the interface between each system's components [17]. The ultimate goal of open system programming is to use one software with all computer systems without incurring further development or porting costs. Development-from-scratch costs are incurred when the whole software for a particular system is rewritten, usually at the request of a company that manufactures the system[86], or when some software functions are added to the originally developed version to enhance its user demand. Porting costs are incurred when parts of the software that handle its interface with the hardware are rewritten to adapt it to the configuration of the host computer system. The previous software issues are most relevant in the design of computer operating systems (OSs). A brief history and design structure of the most successful workstation environment and open system software, UNIX, is presented below [17, 71]. The description of the design structure includes a trace through a UNIX shell command and a presentation of the most common UNIX-managed functions.

2.5.1 UNIX History

UNIX has become the standard workstation OS component and the first open system software to receive wide acclaim from the scientific and

---

[86]For example, IBM developed AIX, a version of the operating system UNIX, to be used in its RISC System/6000 workstations.

programming communities [17]. Developed at Bell Laboratories by Ken Thompson and Dennis Ritchie in the late 1960s [71], UNIX did not gain popularity and market share until, in the 1970s, DARPA[87] funded the University of California at Berkeley to develop a standard UNIX system for government use. The Berkeley UNIX system was supposed to provide networking support for DARPA's Arpanet and local area networks (LANs). Since the Berkeley UNIX system was a government-funded unclassified project, the public had access to it as free software. The workstation companies, in particular Sun Microsystems, Inc., capitalized on that opportunity and used the Berkeley UNIX in their first workstations. Aside from being free, UNIX was chosen as the workstation's OS of choice because it provided the software support for networking, multitasking, and distributed computing, had a small size, and was laid out in a modular, clean design [4, 29, 71].

2.5.2 UNIX Design Structure

As a computer resources and network manager, Berkeley UNIX was designed in two parts, the system programs and the kernel [71].

System Programs

Figure 2.10 illustrates the system programs that constitute the first layer of the OS and handle the interface between the user and the kernel. The user interacts with the computer through this layer by creating an environment where he or she can issue commands to the hardware to execute and receive a response. This environment in UNIX is called a shell. A user can create many shells, and, with the advances in computer/user interface software, several "window"-based interactive software applications have been developed wherein each window is a shell in itself.

---

[87]DARPA stands for Defense Advanced Research Projects Agency.

2.5. UNIX OPERATING SYSTEM

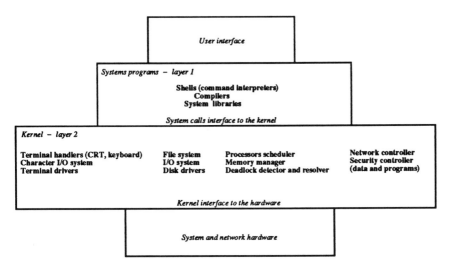

Figure 2.10: UNIX layer structure.

As soon as the user issues a command, it is interpreted and matched within a library of commands. If the library validates the system's acceptance of such a command, a system call is issued to the kernel to access the command's compiled code and to execute it while being assisted by the computer's hardware. UNIX system calls are categorized as process[88] control calls and file and information manipulation calls [71]. Most of the system programs in the shell commands library are written in C, which facilitates moving the software to different hardware platforms and eliminating the hassle of binary code compatibility and portability. If the OS is moved, all that is needed is a C compiler on the host machine to map the C source code of the programs to object binary codes understandable by the machine. (Depending on the hardware platform some software porting might be required.) Source code is usually a synonym for a program written in a high-level language. Object code is usually a synonym for a machine assembly code of a program.

Kernel

After being interpreted by the first layer, system programs are passed as system calls to the kernel, or the second layer, which handles the interface

---

[88] In Peterson and Silberschatz [71], a process is defined as a program in execution.

with the computer's hardware. OS porting is usually performed at this level, if needed. As shown in Figure 2.10, many tasks are performed by the kernel, the first of which is managing the I/O of the computer, in particular, between the processors, the main memory, the hard disk, the monitor, the printer, the mouse, and the keyboard. Other forms of I/O occur at the local or wide area network level. Before discussing the networking capabilities of UNIX, a brief description of the execution of a command in a UNIX-managed system is in order.

2.5.3 Stepping Through a UNIX Command

Suppose that the command requests the user's current directory content to be listed (the *ls* command). The keyboard is used as the standard input device and the monitor as the standard output device. As the command is typed, the shell feeds it to the screen buffers, and the characters of the command are echoed on the monitor. As soon as the carriage return is pressed, the command is interpreted and forwarded to the kernel. The kernel queues the request in the file system, usually a magnetic hard disk, with all the coordinates of the corresponding directory[89] and waits. The disk returns the requested data to the kernel in a buffer. The kernel, correspondingly, transfers the data to the main memory. In turn, main memory feeds it to the screen update buffers, and the directory listing is scanned on the monitor. There are several configurations for establishing the link between the kernel and the I/O devices, a detailed presentation of which can be found in Peterson and Silberschatz [71].

2.5.4 Main UNIX Managed Functions

The main functions handled by UNIX are memory management, multitasking, detecting and resolving deadlocks, security, network communication, and distributed computing.

Memory Management

When executing shell commands, the management of main memory and permanent storage memory is handled by the operating system, especially during program execution. When the data requested by the program are not available in main memory, the OS issues the hardware

---

[89]The directory system in UNIX follows a tree structure, with several tools to protect it and augment it.

2.5. UNIX OPERATING SYSTEM

calls to fetch them and make them available to the running program as efficiently as possible; again, consult [71] for more details.

Multitasking

While waiting for data, the kernel facilitates the multitasking operation of the workstation by efficiently scheduling the processor or processors for different user-requested tasks. Multitasking is performed when a single user is running more than one application in one or several shells, and the OS has to swap the applications' processes in and out of main memory for the processor or processors to complete or update their execution. For each of these applications, the processor must be scheduled for a certain period of time, depending on the priority of the application, its size, and the main memory it requires.

Deadlocks Handling and Security

Other computer managerial issues face the OS, the most important of which are detecting and resolving deadlocks[90] and protecting the overall computer system from being infiltrated by computer pirates. Especially with the rapid growth of computer networks, UNIX-provided security has become a priority for privacy and data protection requirements.

Network Communication and Distributed Computing

UNIX facilitates the setup of local area networks, or workstation-distributed computing environments, because it has the kernel communication routines that enable the launching of different processes[91] on different workstations, with each process reporting to the same originating machine. Connected on a Ethernet or a token ring, the workstations can communicate with each other over the network's fiber-optic lines and with other networks via communication protocol routines performed by the kernel of each network. Usually a network has a file server that acts as the second OS of all network-connected machines and as their gateway to the rest of the world. (For more details regarding distributed

---

[90] A deadlock occurs when a process, called P1, requests a hardware resource in order to terminate, and that resource is held up by a process, called P2, and P2 requires P1 to finish in order to terminate.

[91] In the literature, these processes are referred to as remote procedure calls or remote login sessions.

computing and its issues, see Bertsekas and Tsitsiklis [6] and Peterson and Silberschatz [71].)

2.5.5 Concluding Remarks

Since the American Telegraph and Telephone corporation (AT&T) developed the first version of UNIX in the late 1960s, several other compatible versions from AT&T and other companies have emerged. DEC developed its own Ultrix OS for its DECstations; IBM developed AIX for the RISC System/6000 machines; Sun Microsystems, Inc., developed SunOS for its SUN and SPARC series of workstations; Microsoft developed Xenix for PCs; and several other organizations like Uniforum and Usenix have helped in marketing the UNIX OS to different hardware platforms other than the workstation's. However, the diversification of UNIX development houses created instability in a market that is searching for a standard, off-the-shelf version of the operating system [32]. To deal with the diversity, Posix was developed as a platform of technical standards for users and programmers, and it provided the features and the characteristics the UNIX operating system should have to become universally accepted, across all available hardware platforms [32].

3

Workstation Supply Models

This chapter describes models for the supply of workstation components and workstation assembly. Section 3.1 reviews the simulation approach used. Section 3.2 presents the discrete event simulation supply models of microprocessors and DRAMs, section 3.3 magnetic hard disks, section 3.4 color CRT displays, section 3.5 UNIX operating systems, and section 3.6 the linear workstation assembly process model. Several components attributes and trends will be presented, and the definitions, terminology, and concepts detailed in Chapter 2 will be referred to as this chapter progresses.

3.1 Simulation Overview

Before presenting the supply models, a discussion of the simulation modeling approach adopted in this book is presented in subsection 3.1.1, and a diagrammatic approach to communicating the model's structure is given in subsection 3.1.2.

3.1.1 The Simulation Modeling Approach

Implementing an idea or a system design involves labor and material costs and, if the system undergoes several design or structural changes, these costs can be incurred several times over. Simulation is a tool to model the behavior of the system, experiment with it, and tune it to obtain the desired behavior without incurring the material costs. Certainly there are labor costs incurred in developing the simulator, but these costs are offset by the reduction of risk and unpredictability associated with the behavior of any new design or any modification to an old design [36, 57]. Once the system's inputs, outputs, and its inner working characteristics have been defined, the effects of any change to the system can be traced through the mathematical equations in a way understandable by the simulation tool, and the consequences of that change can be analyzed.

Computer simulation, by its nature, involves a certain amount of art in creating the model. Once created, though, the model can be systematically used to increase our understanding of the actual system's dynamics through simulation experiments. In itself, simulation is not an optimization tool [57], but several simulation techniques can be incorporated into optimization mathematical models. Moreover, the simulation modeler's job does not stop when the development is completed, because validating the model's behavior and verifying its results are as important as developing the simulation model itself. Further discussion of these model verification and validation issues, however, is deferred until Chapter 4.

Computers are efficient simulation tools. During their emergence in the early 1950s, analyses of model designs were done via simulation by mapping the model designs onto the assembly language of the computers or onto a high-level programming language like Fortran and measuring their performance. Not until Geoffrey Gordon[1] introduced the General Purpose Simulation System (GPSS) in 1961 did computer simulation become a common tool, used by large corporations to answer "What if...?" questions without incurring millions of dollars in costs [57].

The three major types of computer simulations are continuous simulation, Monte Carlo simulation, and discrete event simulation [57]. In continuous simulations, the events are generated and evaluated continuously in time. In Monte Carlo simulations, events are generated randomly in time, according to some predefined probability distribution. In discrete event simulations, events are separated by blocks of time.

Since the market data collected are chiefly available on a yearly basis—that is, discrete data—a discrete event simulation approach was chosen to model the dynamics of the supply of workstation components in this book. The methodology is deterministic[2], and the blocks of time

---
[1]Gordon developed GPSS at IBM.
[2]A deterministic system has no random values as input or output and, for each set

3.1. SIMULATION OVERVIEW

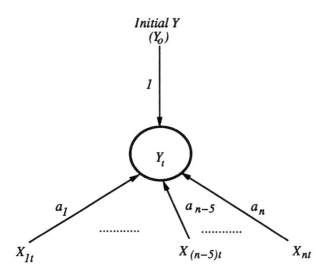

Figure 3.1: Illustration of a relational diagram.

elapsed between each event are equal to one year. As the simulation evolves from year to year, the technological attributes of the components change in response to various technology-driving trends. The relationships between the component attributes and their technology-driving trends are presented in a set of relational diagrams. The relational diagram is the diagrammatic approach to displaying the model relations adopted here to help communicate the content of the component supply models.

3.1.2 Relational Diagrams and Attributes

Figure 3.1 is an illustration of a relational diagram. In Figure 3.1, the state, or value, of an attribute at time t is represented as Y_t and its initial state as Y_0. The index 0 indicates the year the simulation starts and the index t indicates a discrete year between the initial and the final years of the simulation period. Since the data are available on an annual basis, the simulation period of the study is given in number of years from the initial year, and the time index t is incremented by 1 every iteration

---

of input data, there exists one output.

of the model. The figure illustrates n technology-driving trends (TDTs), the relative changes of which are denoted by X_{nt}s. Each technology-driving trend may acquire a new value every iteration, depending on the equation modeling its behavior. Only the relative change in the technology-driving trend's value at t (TDT_t) with respect to its initial value at t_0 (TDT_0) is needed to capture how it affects the attribute Y_t. Hence, each X_{nt} term is expressed as the ratio of the current and initial values of the technology-driving trend it represents:

$$X_{nt} = \frac{TDT_{nt}}{TDT_{n0}}. \qquad (3.1)$$

As illustrated in Figure 3.1, the attribute Y_t is encircled in the center of the relational diagram, with its initial value Y_0 in the upper half of the figure. The relative changes of the technology-driving trends' values, the X_{nt}s, are shown in the lower half of the figure. Y_0 and the X_{nt}s are connected to the encircled Y_t by directed edges, each with an edge weight label. The weights on each X_{nt} edge, the a_ns, are the exponents of these elements in the equation representing the behavior of the attribute. While tuning the values of the attribute's equation to match related actual market data, each a is chosen to reflect the degree to which each technology-driving trend affects, relatively, the attribute. If the absolute value of a_n is larger than the absolute value of $a_{n-\alpha}$ ($|a_n| > |a_{n-\alpha}|$), the effect of a change in TDT_{nt} is considered more important to the overall value of the attribute than a change in $TDT_{(n-\alpha)t}$. If a is positive, an increase (or a decrease) in the technology-driving trend's value increases (or decreases) the value of the attribute; if a is negative, an increase (or a decrease) in the trend's value decreases (or increases) the value of the attribute. The effect of the relative changes in the technology-driving trends' values on the attribute in the relational diagram can be expressed mathematically as:

$$\begin{aligned} Y_t &= Y_0^{a_0} * X_{1t}^{a_1} * X_{2t}^{a_2} * \ldots * X_{nt}^{a_n} & (3.2) \\ &= Y_0^1 * \left(\frac{TDT_{1t}}{TDT_{10}}\right)^{a_1} * \left(\frac{TDT_{2t}}{TDT_{20}}\right)^{a_2} * \ldots * \left(\frac{TDT_{nt}}{TDT_{n0}}\right)^{a_n} & (3.3) \end{aligned}$$

If the partial derivative[3] of Y_t is computed with respect to one of the time-varying technology-driving trends, TDT_{1t} for example, equation

---

[3]The actual technology-driving trends data are chiefly available on a yearly basis—that is, discrete data. However, continuous lines can be drawn through the discrete data, continuous functions can be used to model the behavior of the technology-driving trends over time, and derivatives of the continuous functions can be computed.

3.1. SIMULATION OVERVIEW

(3.3) will be equal[4] to:

$$\frac{\partial Y_t}{\partial TDT_{1t}} = Y_0 * a_1 * \frac{TDT_{1t}^{a_1-1}}{TDT_{10}^{a_1}} * \left(\frac{TDT_{2t}}{TDT_{20}}\right)^{a_2} * \ldots * \left(\frac{TDT_{nt}}{TDT_{n0}}\right)^{a_n}. \quad (3.4)$$

If the positions of Y_0 and a_1 of the right-hand side of equation (3.4) are switched, it will look as follows:

$$\frac{\partial Y_t}{\partial TDT_{1t}} = a_1 * Y_0 * \frac{TDT_{1t}^{a_1-1}}{TDT_{10}^{a_1}} * \left(\frac{TDT_{2t}}{TDT_{20}}\right)^{a_2} * \ldots * \left(\frac{TDT_{nt}}{TDT_{n0}}\right)^{a_n}. \quad (3.5)$$

Moreover, when comparing equation (3.5) with equation (3.3), it is apparent the right-hand side of equation (3.5) can be written as:

$$\frac{\partial Y_t}{\partial TDT_{1t}} = a_1 * \frac{Y_t}{TDT_{1t}}. \quad (3.6)$$

By dividing both the left and the right sides of equation (3.6) by the ratio $\frac{Y_t}{TDT_{1t}}$, equation (3.6) becomes:

$$\frac{\frac{\partial Y_t}{\partial TDT_{1t}}}{\frac{Y_t}{TDT_{1t}}} = \frac{\frac{\partial Y_t}{Y_t}}{\frac{\partial TDT_{1t}}{TDT_{1t}}} = a_1. \quad (3.7)$$

The middle fraction of equation (3.7) can be approximated by the ratio of the fractional change of Y_t and the fractional change of TDT_{1t}:

$$\frac{\frac{\partial Y_t}{Y_t}}{\frac{\partial TDT_{1t}}{TDT_{1t}}} \simeq \frac{\frac{\Delta Y_t}{Y_t}}{\frac{\Delta TDT_{1t}}{TDT_{1t}}} \simeq a_1. \quad (3.8)$$

From equation (3.8), it is apparent that the fractional change in Y_t—i.e. the ratio $\frac{\Delta Y_t}{Y_t}$—can be approximated by the product of the technology-driving trend's fractional change and its exponent in equation (3.3):

$$\frac{\Delta Y_t}{Y_t} \simeq a_1 * \frac{\Delta TDT_{1t}}{TDT_{1t}}. \quad (3.9)$$

Thus, if the value of the technology-driving trend TDT_{1t} were to change by 10 percent, the value of the attribute Y_t changes by approximately 10 percent multiplied by the value of the technology-driving trend's exponent, a_1.

---

[4]The value of a_0 in equation (3.2) is equal to 1 because Y_0 represents the value of the attribute at the first year of the study period, t_0 (if the values of all the X_{nt}s affecting Y_t are equal to 1, $Y_t = Y_0 = Y_0^{a_0}$; i.e., $a_0 = 1$).

Relational Diagrams and Attributes: Example

It was pointed out in Chapter 2 that the CPU's feature size and die area affect its speed[5] attribute. Hence, if speed is denoted as Y_t in equation (3.3) and feature size and die area are denoted as TDT_{1t} and TDT_{2t}, then the speed attribute equation might be expressed mathematically as:

$$SPEED_t = SPEED_0^{a_0} * \left(\frac{FeatureSize_t}{FeatureSize_0}\right)^{a_1} * \left(\frac{DieArea_t}{DieArea_0}\right)^{a_2}. \quad (3.10)$$

The value of a_0 in equation (3.10) is equal to 1 because $SPEED_0$ represents the value of $SPEED_t$ in the first year of the study period, t_0. The values of a_1 and a_2 are set while tuning the values of equation (3.10) to match actual CPU speeds data over a certain period of study; the absolute value of a_2 ($|a_2|$) ought to be larger than the absolute value of a_1 ($|a_1|$) because die area has a relatively stronger influence on the CPU speed than does feature size. For example's sake, let us assume that the values of a_1 and a_2 are, respectively, set to -0.3 and 1 during the tuning process; the CPU speed attribute equation will look as follows:

$$SPEED_t = SPEED_0^{1} * \left(\frac{FeatureSize_t}{FeatureSize_0}\right)^{-0.3} * \left(\frac{DieArea_t}{DieArea_0}\right)^{1}. \quad (3.11)$$

Equation (3.9) shows that a 5 percent decrease in the feature size is reflected by an approximate decrease of -1.5 percent (-0.3 * 5 percent = -1.5 percent), or increase of 1.5 percent, in the CPU speed. On the other hand, a 5 percent increase in the feature size creates an approximate increase of -1.5 percent (-0.3 * 5 percent = -1.5 percent), or decrease of 1.5 percent, in the CPU speed, assuming equation (3.11) is correct.

Equation (3.9) shows also that a 5 percent increase in the die area is reflected in an approximate increase of 5 percent in the CPU speed, and similarly for a die area decrease. Note that the approximation in equation (3.9) carries with it a canceling effect if an increase (or a decrease) occurs in both the feature size and the die area as a result of their exponents' opposite signs; for example, an increase (or a decrease) of 5 percent in the feature size can be, approximately, canceled out by an increase (or a decrease) of 1.5 percent in the die area, and vice versa.

---

[5] For example's sake, only two technology-driving trends are considered. For a complete development of the CPU speed attribute model, refer to section 3.2.

3.2 ICs Supply Model: Microprocessors and DRAMs

For the past four decades, capital intensity[6] and short product life[7] have been the chief characteristics of the semiconductor industry. By the year 2000, a new advanced semiconductor factory will probably cost about $2 billion ($10^9$), and the United States will only be able to afford about 10 such factories [69]. In 1987 dollars, a fabrication laboratory cost approximately $100 million ($10^6$), and required a 3 percent to 6 percent forecasted market share before it could be built [106]. Yet, despite the large capital outlays necessary to build a semiconductor factory, each semiconductor product may average no more than six months to two years of profitable sales before being obsoleted by new and enhanced products!

The main semiconductor components on the market today are microprocessors, microcontrollers, and memories. Many of these products have several generations of components already obsolete before them, and most of them have one or two future versions under design and testing in the manufacturing laboratories. In this book, the components examined are microprocessors and memories—in particular, CISC and RISC microprocessors and DRAMs. The CMOS semiconductor technology is given the greatest attention here.

This section is organized as follows: subsection 3.2.1 presents historical data on the physical characteristics trends of ICs; subsection 3.2.2 presents historical data on capabilities and price trends of ICs; subsection 3.2.3 presents the assumptions and the terminology used in the development of the supply model of ICs, and subsections 3.2.4 through 3.2.7 present the mathematical formulation of the supply model of CPUs and DRAMs.

3.2.1 Historical Data on the Physical Characteristics of ICs

If a company manufacturing ICs is to survive in this competitive and diversified market, its production must be characterized by high product

---

[6] In this context, capital intensity is synonymous with the large amount of capital required to set up an IC manufacturing plant.

[7] Short product life is synonymous with fast obsolescence rate.

yields. This criterion can only be satisfied with a clean manufacturing environment, constant attention to the latest technological developments, and learning[8]. Through learning, acquired techniques are transferred to new generation of products without reinventing the wheel.

Today, the Japanese fabrication laboratories are the most sophisticated and cleanest in the world. The position of the Japanese ICs manufacturing firms on the learning curves and their clean environments have enabled them to achieve higher yields than their U.S. and European counterparts. This allowed them to launch a DRAM dumping war in the 1980s and capture almost 90 percent of the world DRAM market [26].

As described in Chapter 2, the current technology for manufacturing ICs is very intricate, involving several masking levels and processing steps, thereby making the ICs quality and reliability very sensitive to a whole chain of decisions during the production process. Adding to the production processes' intricacy are the ICs' functionalities and complexity. What follows is a presentation of ICs production trends[9], including the increase of the die area, the reduction of the feature size, changes in the chip's configuration, the increase of the number of critical masks, and the increase of the silicon wafer diameter.

ICs Die Size Trends

Die size[10] has steadily increased [49]. Figure 3.2[11] is a log-linear graph that illustrates the historical trend of increasing die areas for processors, ASICs, and memories, where the vertical axis indicates the base 10 logarithm of the chip's area[12] in square millimeters and the horizontal axis indicates the year of the ICs market introduction. Die areas of DRAMs and of Motorola's and Intel's CISC microprocessors can be followed in Figure 3.2. For instance, the die area of a 4 megabit DRAM, introduced in 1988, was close to 80 mm^2, and the die area of the Intel i80386 microprocessor, introduced in 1986, was close to 113 mm^2. Using the upper

---

[8]Learning is dependent on good documentation of the manufacturing processes.

[9]Production trends refer to manufacturing trends and physical characteristics of the components. In this book, they are called technology-driving trends.

[10]Die size and die area are synonymous in this context.

[11]G. Kötzle, "VLSI Technology Trends," **IEEE Proceedings, COMPEURO'89: VLSI and Computer Peripherals**, ©1989 IEEE. Reprinted with permission of IEEE Press.

[12]100 mm^2 = 1 cm^2.

3.2. ICS SUPPLY MODEL: MICROPROCESSORS AND DRAMS

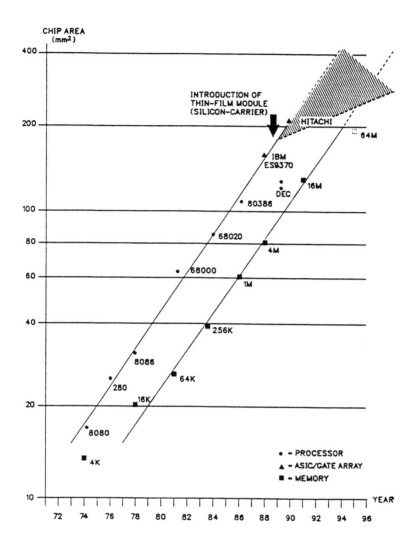

Figure 3.2: Die sizes of CISC CPUs and DRAMs versus time. Source: [49]. ©1989 IEEE.

line shown in Figure 3.2 to relate the die area of a CISC microprocessor to its year of manufacture, the following equation results:

$$CISCDA_t = 10^{-0.1+0.0753*(t-1984)} \ (cm^2) \qquad (3.12)$$

where $CISCDA_t$ represents the CISC CPU die area in cm^2, and t represents the year the die area is calculated. Similarly, the lower line of Figure 3.2 can be used to relate the die area of a DRAM to its year of manufacture. This equation can be written as:

$$DRAMDA_t = 10^{-0.4+0.0753*(t-1984)} \ (cm^2). \qquad (3.13)$$

The die area trends presented in equations (3.12) and (3.13) will taper off in the future because of the manufacturing difficulties and the low yields associated with large area dies [33, 49, 72]. While the equations' values capture past trends, they do not project, with absolute certainty, future die areas. Nevertheless, they can be used to illustrate a "possible" future behavior of IC attributes. What follows is a presentation of the die yield as the main factor inhibiting the die area increase.

ICs Die Yields

Die yields data corresponding to different die areas was presented in Chapter 2 of Hennessy and Patterson [33]. The data, shown in Table 3.1, indicate that an increase in the die area induces a decrease in the die yield and, consequently, an increase in the cost of the die. A larger die has a larger area exposed to dust particles and manufacturing defects. Therefore, clean production environments and continuous production processes enhancements are essential for high yields. Since die yields are so crucial to the economics of producing ICs, the supply model will incorporate several factors affecting the die yield, including the die area, and generate the die yields for the corresponding die areas. (The model's die yield results will be presented in subsection 4.3.1 and compared to the data provided in Table 3.1.)

ICs Feature Size Trend

The minimum ICs feature size has been decreasing since the early 1960s [94]. A smaller feature size is usually accompanied by a faster die operational speed, a higher die functionality, and an increase in its manufacturing complexity and testing time. The historical trend of the minimum

Table 3.1: Die areas and their actual die yields. Source: [33].

ICs: Die Yields	
Die Area (cm^2)	Die Yield (Actual Data) (%)
0.0625	78
0.2601	46
0.5776	22
1.0404	10
1.6129	5
2.3104	3
3.1684	2
4.1209	1

feature size is presented in Figure 3.3[13], a log-linear graph where the vertical axis indicates the base 10 logarithm of the feature size in microns and the horizontal axis indicates the time. On the graph, the feature size capabilities of different types of photoresists, lithography methods, and pattern etching technologies are included, each with time intervals representing the time of introduction and the time of obsolescence. For instance, the negative photoresist had been used with contact printing lithography and wet pattern etching technology from the early 1960s through the late 1970s. The line shown in Figure 3.3 can be expressed as the following equation:

$$FS_t = 10^{1.4-0.055*(t-1960)} \; (micron) \qquad (3.14)$$

where FS_t represents the minimum feature size in microns, and t represents the year in which the feature size is calculated.

ICs Die Configuration Trend

Another factor that has an important effect on a die's functionality is its overall configuration. For example, new generation microprocessors

---
[13] Al Tasch, **Class Notes for EE396K, MOS-IC Process Integration**, ©1990 Al Tasch. Reprinted with permission of Al Tasch.

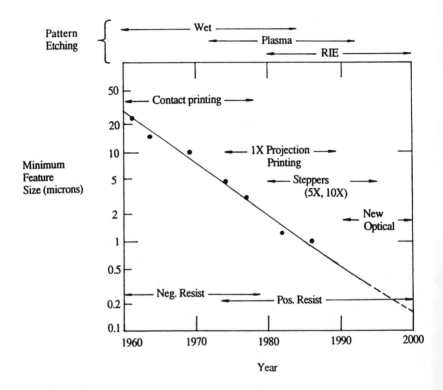

Figure 3.3: Feature size versus time. Source: [94]. ©1990 Al Tasch.

are configured with an integer unit (IU), a floating point unit (FPU), and a memory management unit (MMU) on one single die, reducing the processing board's interconnection delays and improving the throughput. Other configurations dedicate a die to each of the IU, FPU, and MMU functions by using an enhanced packaging technology. In the 1970s, the packaging technology had to handle the high power consumption of bipolar ICs; however, the current semiconductor CMOS technology is characterized by very low power consumption. As a result, newer and more sophisticated packaging techniques have been developed, each with its physical limitations on the number of pins[14] the package can have and, correspondingly, the maximum number of watts[15] it can consume [101]. Nevertheless, the maximum number of pins per CMOS package is far greater than for an equal area bipolar package when both chips are designed to operate at room temperatures.

Packaging is itself a highly developed field. Several new packaging technologies are discussed in [49, 67, 74, 97, 101]. One of the most promising new approaches is to distribute the dies on a silicon-based thin-film package, reducing communication and interconnection delays and allowing the dies' operational speeds to reach their respective limits without compromising performance with power consumption issues [49, 74].

ICs Production: Number of Critical Masks Trend

As presented in section 2.2, during the production of an IC, several photoresist masks are used in the process of etching the circuit configuration and doping the corresponding etched areas with the specified impurities. The number of these masks depends on the functionality of the IC and the manufacturer's approach to achieving that functionality. Of these masks, a few are critical to the production of a good die and, consequently, its cost. The current average number of masks for most CMOS ICs is within the 14 to 20 range [18], and the number is projected to increase to 25 by the year 2000. The current average number of critical masks is within the 2 to 4 range [18, 33], or about 25 percent of the total number of masks. If the number of critical masks stays constant as a fraction of the total number of masks, then, by the year 2000, it will

---

[14]Package pins are leads connecting the die to its corresponding socket on the processing or memory board.

[15]Generated or consumed electric power is measured in watts.

grow to six to eight in number. If we express this trend mathematically, the number of critical masks can be captured by the following equation:

$$CM_t = 25\% * \lfloor 14.5 * 2^{0.08*(t-1990)} \rfloor \qquad (3.15)$$

where CM_t represents the number of critical masks, and t represents the year in which the number of critical masks is calculated. Using equation (3.15), the number of critical masks in 1980 is close to 2, a number that comports with actual data provided in [18, 33].

ICs Production: Silicon Wafer Diameter Trend

The trend in silicon wafer diameter, shown in Figure 3.4[16], has been toward increased size since the early 1960s [94]. The larger the wafer diameter is, the larger the number of dies per wafer and the cheaper the cost per die. The cost of the wafer increases as its area increases; however, the increase in the overall number of dies that can be produced from it offsets the wafer's cost increase. The log-linear graph of the historical trend of the wafer diameter, presented in Figure 3.4, has a vertical axis measuring the base 10 logarithm of the wafer diameter in millimeters and a horizontal axis indicating time. A line drawn through the points in Figure 3.4 to relate wafer diameter to time results in the following equation:

$$WD_t = 10^{1.48+0.0252*(t-1960)} \ (mm) \qquad (3.16)$$

where WD_t represents the silicon wafer diameter in millimeters, and t represents the year in which the wafer diameter is calculated.

Closure

When taken together, the historical trends equations—die area, feature size, number of critical masks, silicon wafer diameter—presented in the current subsection form a set that captures the ICs technology-driving trends as a function of time. These equations will be used as a part of the ICs supply model, to be presented later in this section.

---

[16]Data from 1960 to 1982: S.M. Sze, **VLSI Technology**, ©1983 McGraw-Hill. Reprinted with permission of McGraw-Hill Book Company. Data from 1983 to date: Al Tasch, **Class Notes for EE396K, MOS-IC Process Integration**, ©1990 Al Tasch. Reprinted with permission of Al Tasch.

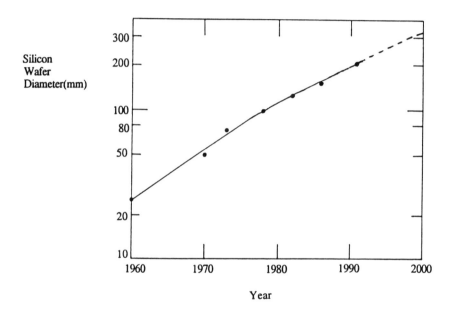

Figure 3.4: Silicon wafer diameter versus time. Sources: [89, 94]. ©1983 McGraw-Hill, ©1990 Al Tasch.

3.2.2 Historical Data on ICs Capabilities and Price Trends

Actual data on ICs capabilities and prices were collected from several sources, including newspapers, magazines, journals, computer news network colleagues, and companies' data manuals. What follows are tabularized and graphed actual data of CISC and RISC CPUs and DRAMs.

CPUs Data

The following paragraphs present actual CISC and RISC CPUs price, performance, and physical dimension data, including CPU speed, MIPS, IPC[17], and die area. The MIPS rating can be used as a *relative performance measure only to gain insight into the operational throughput of the CPU itself and not the whole computer system.*

Actual CISC CPUs Data

Table 3.2 provides actual Intel and Motorola CISC CPUs data, including the CPU introduction date, its model number, its operational speed, its MIPS rating, its number of instructions per cycle (IPC) value, its die[18] area in (mils x mils) and cm^2, and, finally, its speed per square centimeter rating. (Unavailable data are listed on the table as n.a.) Some of the listed processors perform integer operations only, and coprocessors perform the floating point and memory management operations.

The CPU's throughput is a function of its speed and its architecture. The speed reflects the sophistication of the manufacturing process and the feature size at which a CPU is manufactured. The architecture encompasses the instruction set used, CISC or RISC, and the number of instructions per cycle. The IPC values on Table 3.2 are obtained by dividing the MIPS rating by the speed of the corresponding CPU, and the speed per square centimeter values are obtained by dividing the speed of the corresponding CPU by its die area.

Table 3.3 provides actual Intel CISC CPUs prices and price per MIPS data over the 1979–1991 period. There is a difference, however, between the list price and the direct cost of producing an IC. The list price of an

---

[17]IPC is an acronym for the number of Instructions Per Cycle.
[18]1,000 mils = 1 inch = 2.54 cm.

Table 3.2: Actual market data of Intel and Motorola CISC CPUs. Sources: Intel data [42, 43, 49], Motorola data [62, 72, 85].

ICs: CISC CPUs							
Year	CPU Model #	Speed (MHz)	MIPS	IPC	Die Area (mils x mils)	Die Area (cm^2)	Speed per cm^2
1978	i8086	8	n.a.	n.a.	n.a.	0.28	28.6
1979	i8088	8	0.33	0.04	n.a.	n.a.	n.a.
1982	M68000	10	0.5	0.05	215x239	0.33	30.3
1982	i80286	16	2	0.125	n.a.	n.a.	n.a.
1983	M68010	10	0.6	0.06	239x267	0.41	24.4
1985	M68020	16.6	2	0.12	271x278	0.49	33.8
1985	i80386	16	3	0.187	n.a.	1.13	14.15
1987	M68020	20	3	0.15	271x278	0.49	40.8
1987	M68030	25	4	0.16	330x379	0.80	31.2
1987	i80386	20	4.3	0.215	n.a.	1.13	17.7
1988	i80386	25	6.1	0.244	n.a.	1.13	22.1
1988	i80386	33	8.3	0.25	n.a.	1.13	29.2
1989	i80486	25	11.4	0.45	414x619	1.65	15.2
1990	M68040	33	20	0.606	490x460	1.45	22.75
1990	i80486	33	15.2	0.46	414x619	1.65	20
1990	i80486	50	25	0.5	414x619	1.65	30.3
1991	i80486	66	35	0.53	414x619	1.65	40

Table 3.3: Actual prices and price per MIPS market data of Intel CISC CPUs. Sources: [42, 43].

\multicolumn{4}{c}{ICs: CISC CPUs}			
Year	Model #-Speed (MHz)	Price (Actual Data) ($)	Price per MIPS (Actual Data) ($)
1979	i8088-8	360	1091
1982	i80286-16	360	180
1985	i80386-16	299	99.6
1989	i80486DX-25	700	61.4
1991	i8086-8	1.5	n.a.
1991	i80286-16	7	3.5
1991	i80386DX-33	166	20
1991	i80486DX-66	588	16.8
1991	i80486SX-25	258	22.6

IC is usually equal to [33]:

$$List\ Price = \frac{Cost * (1 + Direct\ Cost\ \%)}{(1 - Gross\ Margin\ \%) * (1 - Average\ Discount\ \%)}\ (\$) \quad (3.17)$$

where the direct cost includes labor and product overhead costs; the gross margin incorporates the overall pretax profits, taxes, production costs (like R&D), marketing, sales, maintenance, loans interest payments, and rental payments [33]; and the average discount is used as an incentive for large volume buyers. The distinction between list price and direct cost is drawn because the supply models to be presented later will provide only the direct costs of the components, not the list price. To obtain the list price, a markup of up to 200 percent to 300 percent is necessary, depending on the ICs manufacturer and the product [33]. Each company might have different percentages for the direct costs, the gross margin, and the average discount, and the values of their percentages depend on the product's attributes and quality, the market competition, and the overall market demand for similar products.

3.2. ICS SUPPLY MODEL: MICROPROCESSORS AND DRAMS

Actual RISC CPUs Data

Table 3.4 presents speeds, MIPS ratings, and IPCs of the Hewlett-Packard Precision Architecture (HP-PA) and the Sun Microsystems Scalable Processor Architecture (SPARC) RISC machines. No CPU price data were available, and most of the die sizes are still considered as proprietary company information. As far as the IPC values are concerned, it is apparent that the RISC IPCs are larger than the CISC (for more details refer back to section 2.2).

Actual Data for CISC and RISC CPUs Instructions per Cycle

Log-linear graphs of Intel and Motorola CISC CPU IPCs, and of HP and Sun RISC CPU IPCs are presented in Figures 3.5 and 3.6, respectively, where each vertical axis indicates the base 10 logarithm of the IPC of the corresponding processor and each horizontal axis indicates the time. The CISC IPC trend in Figure 3.5 can be expressed by the following equation:

$$CIPC_t = 0.05 * 10^{0.11*(t-1982)} \qquad (3.18)$$

and the RISC IPC trend in Figure 3.6 can be expressed by the following equation:

$$RIPC_t = 0.6 * 10^{0.063*(t-1987)}. \qquad (3.19)$$

If extrapolated, the trend lines suggest that an ever-increasing IPC is possible over time. In fact, the IPC may be limited to upper bound values in the 3 to 4 range, which might be reached by—if not before—the year 2000 [72, 99]. The limit is set by the architectural designs and the frequency at which a branch instruction is executed. The architectural designs limitations were discussed in section 2.2. The branching limitations are related to the maximum number of instructions that can be executed in one clock cycle.

Each processor's assembly language has a set of branch instructions which are used to move around the program counter[19] in the object code during execution. The program counter's branch destination depends on the type of the branch and the logical condition, if any, that the branch instruction must satisfy before being acknowledged. It has been estimated that there are three to four instructions between two branch

---

[19] A program counter is the processor's hardware pointer to the program being executed.

Table 3.4: Actual performance data of HP-PA and Sun SPARC RISC machines. Sources: HP data [7, 22, 38, 52, 54, 91, 105], Sun data [85, 87].

ICs: RISC Machines

Year	System	Speed (MHz)	MIPS	IPC	Die Area (mils x mils)	Die Area (cm^2)	Speed per cm^2
1987	HP9000/825	12.5	9	0.72	n.a.	n.a.	n.a.
1987	Sun 4/260	16.67	10	0.599	n.a.	n.a.	n.a.
1988	HP9000/835	15	14	0.93	n.a.	n.a.	n.a.
1988	Sun 4/110	14.28	7	0.49	n.a.	n.a.	n.a.
1989	HP9000/845	30	22	0.73	551x551	1.96	15.3
1989	Sparcstation1	20	12.5	0.62	n.a.	n.a.	n.a.
1990	Sparcstation1+	25	15.8	0.63	n.a.	n.a.	n.a.
1990	Sparcstation2	40	28.5	0.71	n.a.	n.a.	n.a.
1991	HP9000/720	50	59	1.14	551x551	1.96	25.5
1991	HP9000/730	66	76	1.15	551x551	1.96	33.67

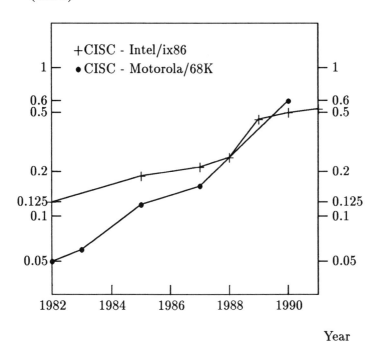

Figure 3.5: Actual data on the number of instructions per cycle for Intel and Motorola CISC CPUs. Sources: Intel data [42, 43, 49], Motorola data [62, 72, 85].

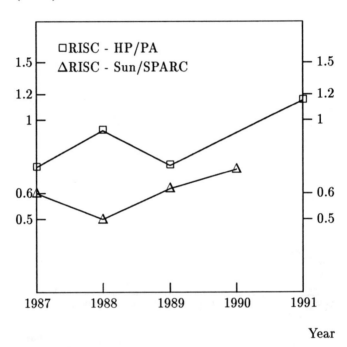

Figure 3.6: Actual data on the number of instructions per cycle for HP-PA and Sun SPARC RISC CPUs. Sources: HP data [7, 22, 38, 52, 54, 91, 105], Sun data [85, 87].

Table 3.5: Actual market data on DRAM die sizes, capacities, and densities. Source: [49].

ICs: DRAMs				
Year	DRAM (bits)	Die Area (cm^2)	Mb/cm^2	MB/cm^2
1978	16K	0.2	0.08	0.01
1981	64K	0.24	0.27	0.034
1984	256K	0.4	0.64	0.08
1986	1M	0.6	1.67	0.21
1988	4M	0.8	5	0.625
1991	16M	1.34	11.94	1.49

instructions, so that up to four instructions might be executed in one cycle (IPC = 4) before the branch occurs and another set of instructions can be executed [72]. This practical limitation in IPCs needs to be kept in mind, since it will influence future projected trends of CPU MIPS or any other CPU performance rating.

DRAMs Data

Table 3.5 provides actual DRAM data. Shown for each DRAM are its capacity in bits, its introduction date, its die size in cm^2, its density in number of megabits per square centimeter (Mb/cm^2), and its number of megabytes per square centimeter (MB/cm^2). The Mb/cm^2 data were obtained by dividing the K-indexed capacity values by the product of the corresponding die area and the number 1,000, and the M-indexed capacity values by the corresponding die area. The MB/cm^2 was obtained by dividing the Mb/cm^2 values[20] by 8.

Actual Motorola DRAM prices and price per megabyte data are presented in Table 3.6. Since the mid-1980s, DRAM prices decreased much faster than the manufacturers expected due to fierce Japanese competition [26]. Again, it is emphasized that the prices shown are list prices, not direct costs, and are marked up over costs by as much as 300 percent [33].

---

[20] Eight bits are required to form 1 byte.

Table 3.6: Actual prices and price per megabyte market data of Motorola DRAMs. Source: [63].

\multicolumn{3}{c}{ICs: DRAMs}		
Year	DRAM (bits)	Price (Actual Data) ($)	Price/MB (Actual Data) ($)
---	---	---	---
1984	16K	1.09	545
1984	64K	3.4	425
1984	256K	17.9	559.3
1986	1M	100	800
1988	4M	264	528
1991	16M	329	164.5

3.2.3 Model Assumptions and Terminology

The ICs supply model incorporates equations capturing the attributes of CMOS CPUs and DRAMs. The attributes considered are the costs, speeds, and MIPS of CPUs, and costs and capacities of DRAMs. Before presenting the mathematical formulation of the model, several assumptions are listed. This is followed by a list defining the variables and the parameters used in the model.

Model Assumptions

A relatively short period of study—1980 through 1996—is adopted because of the rapid pace at which the computer industry is changing and the uncertainties associated with it. Choosing a short period makes possible other assumptions listed below:

- Technological trends of the past in overcoming physical manufacturing barriers continue during the period of study.

- Past trends in IC manufacturing yields will continue to increase or, at worst, remain constant.

- Past trends in IC testing will continue to improve to deal with higher density and higher complexity chips, thereby improving the reliability of the ICs.

3.2. ICS SUPPLY MODEL: MICROPROCESSORS AND DRAMS

Definitions and Terminology

In presenting the model's equations, several parameters and variables are introduced, and they are defined as follows:

0	time index for the first year of the study period
t	time index
FS	minimum feature size, in microns
CM	number of critical masks
WD	wafer diameter, in centimeters
WC	wafer cost, in dollars
DA	die area, in cm^2
DPW	number of dies per wafer
$TDPW$	number of test dies per wafer
DY	die yield, in percent
WY	wafer yield, in percent
FTY	final die-test yield, in percent
$DPUA$	number of silicon wafer defects per unit area
$ADTT$	average die test time, in seconds
$TCPH$	die testing cost per hour, in dollars
TC	die testing cost, in dollars
MC	die manufacturing cost, in dollars
PC	die packaging cost, in dollars
BIC	IC burn-in cost, in dollars
$ICTC$	IC total cost, in dollars
$SPEED$	speed of operation of the IC, in megaHertz
$SPEEDPER$	speed of operation of the IC per square centimeter
$CIPC$	number of CISC instructions per CPU clock cycle
$RIPC$	number of RISC instructions per CPU clock cycle
$CMIPS$	number of million CISC instructions per second
$RMIPS$	number of million RISC instructions per second
$MEMORY$	capacity of a DRAM, in megabytes
$MEMPER$	number of DRAM megabytes per square centimeter

The supply model of ICs is formulated as follows: subsection 3.2.4 presents the CISC and RISC CPU speed and MIPS models; subsection 3.2.5 presents the DRAM capacity model; subsection 3.2.6 presents the time behavior of dies of fixed capability (fixed CPU speed and fixed DRAM capacity ICs); and subsection 3.2.7 presents the IC cost model with a description of the die manufacturing cost, the testing cost, the

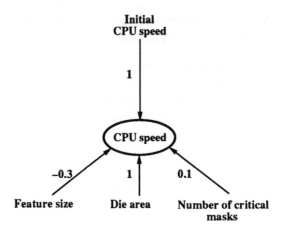

Figure 3.7: Relational diagram of the operational speed of CPUs.

packaging cost, the burn-in cost, and the associated die yields all through the IC manufacturing process.

3.2.4 CPU Speed and MIPS Models

The MIPS rating is used in the model as a relative measure that reflects on the operational speed behavior of the CPU and its architectural enhancements over the years, where the architectural enhancements can be captured by the IPC trends provided earlier. Since there are two different instruction set computers, CISC and RISC, two MIPS equations are formulated accordingly:

$$CMIPS_t = SPEED_t * CIPC_t \qquad (3.20)$$
$$RMIPS_t = SPEED_t * RIPC_t. \qquad (3.21)$$

The relational diagram[21] for the operational speed model is illustrated in Figure 3.7. It relates the CPU speed at t to the initial CPU speed at t_0 and the technology-driving trends of feature size, die area, and number of critical masks:

---

[21] Refer to section 3.1 for a description of the relational diagram symbols and their interpretations.

3.2. ICS SUPPLY MODEL: MICROPROCESSORS AND DRAMS

1. Feature size was chosen as a CPU speed technology-driving trend because a smaller feature size reduces the space between the chip's wires and modules, hence reducing the distance the electrons have to travel and increasing the die speed.

2. Die area was chosen as a CPU speed technology-driving trend because a larger die area makes possible an increase in the functionality of the CPU. Several specialty ASICs that had been implemented outside the CPU, like the memory management unit and the floating point unit, can be included in the larger CPU area, thus eliminating the speed limiting communication outside the CPU and enabling a CPU's speed to reach limits unattainable through the packaging and early manufacturing technologies.

3. Number of critical masks was chosen as a CPU speed technology-driving trend because it reflects the degree of manufacturing sophistication and the level of integration of the CPU production process. Both factors affect the CPU speed in the long run.

The CPU speed model can be expressed mathematically as follows:

$$SPEED_t = SPEED_0 * \left(\frac{FS_t}{FS_0}\right)^{-0.3} * \frac{DA_t}{DA_0} * \left(\frac{CM_t}{CM_0}\right)^{0.1} \quad (MHz). \quad (3.22)$$

It indicates that the CPU speed, at time t, increases as the feature size decreases and as the die area and the number of critical masks increase. The values of the exponents (a_ns) in equation (3.22) were obtained by tuning the model to yield results to match the actual CPU speeds data provided in subsection 3.2.2. All of the $|a_n|$s are ≤ 1, and each reflects an estimate of its relative importance to the behavior of the CPU speed attribute over time. For instance, the exponent of the die area element—$a_2 = 1$—in equation (3.22) is larger than the number of critical masks'—$a_3 = 0.1$—because the die area plays a more important role in determining the CPU speed than the number of critical masks.

If the sign of an exponent (a_n) is positive, an increase (or a decrease) in the corresponding technology-driving trend's value increases (or decreases) the CPU speed; if the sign is negative, an increase (or a decrease) in the trend's value decreases (or increases) the CPU speed. For example, a 10 percent decrease in the feature size leads to an approximate increase of 3 percent in the CPU speed [see equation (3.9)], while a 10 percent increase in the die area leads to an approximate increase of 10 percent in the CPU speed, and a 10 percent increase in the number of critical masks leads to an approximate increase of 1 percent in the CPU speed.

3.2.5 DRAM Capacity Model

The speed of DRAM ICs is critical to a computer system's performance. However, since they are needed in large quantities for the system to have high performance measures, their interconnection and communication delays can offset their operational speed. Fast memory chips with higher bit densities are more desirable than an equal capacity of interconnected chips with a lesser bit density.

The DRAM IC is not as complex as a CPU IC because several processing functions are not implemented onto a single die. Rather, the DRAM has a simple structure formed by a repetition of memory cells, each separated by a pitch usually equal to the feature size. The number of critical masks is essentially equal to that of the CPUs' [18].

The relational diagram of the DRAM capacity model is illustrated in Figure 3.8. It relates the DRAM capacity at t to the initial DRAM capacity at t_0 and the technology-driving trends of feature size, die area, and number of critical masks:

1. Feature size was chosen as a DRAM capacity technology-driving trend because a smaller feature size reduces the space between the memory cells and the space each of them occupies on the die, thus increasing the DRAM capacity.

2. Die area was chosen as a DRAM capacity technology-driving trend because a larger die can incorporate more memory cells and have a larger DRAM capacity.

3. Number of critical masks was chosen as a DRAM capacity technology-driving trend because it reflects more sophisticated DRAM manufacturing processes where more than one memory cell can occupy the same space only one cell occupied in an earlier DRAM layout.

The DRAM capacity model can be expressed mathematically as follows:

$$MEMORY_t = MEMORY_0 * \left(\frac{FS_t}{FS_0}\right)^{-2} * \frac{DA_t}{DA_0} * \left(\frac{CM_t}{CM_0}\right)^2 \ (MB). \quad (3.23)$$

3.2. ICS SUPPLY MODEL: MICROPROCESSORS AND DRAMS

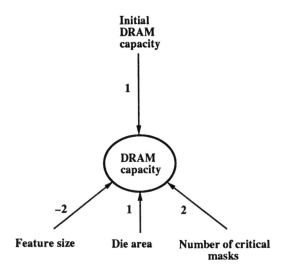

Figure 3.8: Relational diagram of the DRAM capacity.

The DRAM capacity increases as the feature size decreases and as the die area and the number of critical masks increase. The values of the exponents (a_ns) in equation (3.23) were obtained by tuning the model to yield results to match the actual DRAM capacity data provided in subsection 3.2.2. All of the $|a_n|$s are ≥ 1, and each reflects an estimate of its relative importance to the behavior of the DRAM capacity attribute over time. For instance, a 10 percent decrease in the feature size leads to an approximate increase of 20 percent in the DRAM capacity [see equation (3.9)], and similarly for the other factors of equation (3.23).

3.2.6 Die Areas for Fixed Capability ICs

For a fixed functionality or attribute IC (that is, a fixed CPU operational speed in the case of microprocessors, or a fixed DRAM capacity in the case of memories), the die size decreases over time; this is because the minimum feature size decreases and the value of the number of critical masks increases. Decreasing die areas allow increasing die yields and decreasing die costs. The corresponding die area equations for each type of fixed capability IC can be derived from equations (3.22) and (3.23). They are fixed speed CPU and fixed capacity DRAM:

- Fixed Speed CPU

$$DA_t = \frac{Fixed\ SPEED}{SPEEDPER_0} * \left(\frac{FS_t}{FS_0}\right)^{0.3} * \left(\frac{CM_t}{CM_0}\right)^{-0.1} \quad (cm^2) \quad (3.24)$$

where $SPEEDPER_0$ is equal to the initial CPU speed at t_0 ($SPEED_0$) divided by its corresponding die area at t_0 (DA_0).

- Fixed Capacity DRAM

$$DA_t = \frac{Fixed\ DRAM}{MEMPER_0} * \left(\frac{FS_t}{FS_0}\right)^{2} * \left(\frac{CM_t}{CM_0}\right)^{-2} \quad (cm^2) \quad (3.25)$$

where $MEMPER_0$ is equal to the initial DRAM capacity at t_0 ($MEMORY_0$) divided by its corresponding die area at t_0 (DA_0).

3.2.7 IC Cost Model

The total direct manufacturing cost of a packaged and operational IC is equal to:

$$ICTC_t = \frac{MC_t + TC_t + PC_t + BIC_t}{FTY_t} \quad (\$) \quad (3.26)$$

where MC_t represents the IC manufacturing cost, TC_t the testing cost, PC_t the packaging cost, BIC_t the burn-in cost, and FTY_t the final test yield. No IC indirect costs are considered in equation (3.26), as they are assumed as internal company decisions. What follows are descriptions of each of the cost factors of equation (3.26), each with its own mathematical expression.

IC Manufacturing Cost

The die manufacturing cost depends on the silicon wafer cost, the manufacturing die yield, and the number of manufacturable dies per wafer. The manufacturing cost equation is:

$$MC_t = \frac{WC_t}{DPW_t * DY_t} \quad (\$) \quad (3.27)$$

where WC_t is the silicon wafer cost, DPW_t is the number of dies per wafer, and DY_t is the die yield. The paragraphs below discuss these components of equation (3.27) in turn.

3.2. ICS SUPPLY MODEL: MICROPROCESSORS AND DRAMS

Silicon Wafer Cost

Silicon ingot manufacture has improved over the years and, consequently, the quality of the wafers sliced from the ingots has also increased. The yield of good and processable silicon wafers per ingot has reached the 80 percent to 90 percent level during the past decade [18, 33], while, at the same time, the overall cost of a fixed diameter wafer has dropped. However, as the wafer diameters have increased, so have their costs, offsetting some of the cost reductions made possible by improved manufacturing.

The only data available on the wafer cost are found in [33], where a wafer, in 1990, cost from $500 to $550. From equation (3.16), the wafer diameter in 1990 was 16.25 centimeters. Unfortunately, no wafer price data for other years were available, so no price trend could be estimated from historical data. Based on conversations with colleagues, we simply assumed that wafer costs have increased by $20 per year since 1980, which is represented by the equation:

$$WC_t = 350 + (t - 1980) * 20 \ (\$). \tag{3.28}$$

Actual wafer cost data vary from one manufacturer to another, and most companies consider it proprietary information. It is assumed that the $20 per year increase reflects the net effects of diameter increase, inflation, and the improved manufacturing yields.

Number of Dies per Wafer

Since dies are rectangular and wafers are usually round, the partitioning of the wafer results in some round edges that cannot be manufactured into the specified die dimensions. Furthermore, most companies select two or more dies per wafer for testing purposes, which further reduces the area of the wafer used for packable ICs. The number of dies per wafer is expressed as:

$$DPW_t = \frac{\pi * WD_t^2}{4 * DA_t} - \frac{\pi * WD_t}{\sqrt{2 * DA_t}} - TDPW_t \tag{3.29}$$

where the first term is equal to the wafer area divided by the die area, the second term is the wasted round wafer area, and the third term is the number of testing dies per wafer. Usually the number of testing dies per wafer is equal to 2 [33].

Die Yield

By increasing the die area, three consequences ensue: the number of dies per wafer becomes smaller, the cost per die increases [equation (3.27)], and the die yield decreases [33]. Several manufacturing die yield approximations have been formulated [19, 33], and the one chosen here was first presented in Chapter 2 of Hennessy and Patterson [33]:

$$DY_t = WY_t * \left(1 + \frac{DPUA_t * DA_t}{CM_t}\right)^{-CM_t}. \qquad (3.30)$$

Remember that the die yield factor appears in the denominator of equation (3.27), the expression for the manufacturing cost. When averaging the costs of the faulty and good dies, a smaller die yield will lead to a higher cost for each good die. Generally, the die yield has been a major factor in limiting the increase of the average manufacturable die area [33].

Due to the number of masks needed to produce a final chip, the die area could be exposed, at each mask, to etching and dust particles errors and faulty connections. These errors are translated into die defects and cause the elimination of the die from the packaging stage. Equation (3.30) reflects the relative importance of the number of critical masks (CM), the die area (DA), and the number of silicon wafer defects per unit area (DPUA) on the manufacture of dies.

The wafer yield (WY) is included in equation (3.30) because a faulty die could result from a wafer manufacturing defect which was not detected until the die production process. Wafer yields range from 80 percent to 90 percent [18, 33].

Number of Silicon Wafer Defects per Unit Area

The number of silicon wafer defects per unit area is influenced by the feature size's decreasing trend, the increasing trends over time of the IC's number of critical masks and die area, and the manufacture of the silicon wafer. Since each company guards its historical defects per unit area trends as classified information, no actual data were available. Because the improving position of the IC and wafer manufacturers on the learning curves has tended to offset the decrease of the feature size and the increase of the die area and the number of critical IC production masks, the number of silicon wafer defects per unit area was assumed to

3.2. ICS SUPPLY MODEL: MICROPROCESSORS AND DRAMS

stay constant at 2.5 defects per cm^2 during the period of study. This assumption will be modified as necessary when additional data become available.

Manufactured Die Testing Cost

Each manufactured die is put through a series of tests to check its functionality and operational behavior. In some cases, mostly in CPU ICs, more than one function is implemented on a die. So, if one part of the die performing one of its functions is shown by the test to be faulty, this part can be disabled and the CPU sold with the disabled function deleted as one of its attributes. In fact, Intel has done just this with its latest series of i80486 microprocessors. One of the i80486 CPU's functions handles the execution of floating point operations; if the floating point unit (FPU) is operational, the CPU is labeled i80486DX and sold for $588 (see Table 3.3). If it is faulty, the CPU with a disabled FPU is labeled i80486SX and sold for $258. The moral of the story is that, with increasing die areas and die functionalities, it is costly to throw away a die if only a portion of it is faulty and can be disabled.

With each die test performed there is an associated cost. This cost is a function of the testing cost per hour (TCPH) at the manufacturing facility, the average die testing time (ADTT), and the yield of the dies that pass the test (DY). The testing cost equation is:

$$TC_t = \frac{TCPH_t * ADTT_t}{DY_t * 3600} \ (\$) \qquad (3.31)$$

where the testing cost per hour depends on the manufacturer and the testing equipment used. The following paragraphs explain each of these factors.

Die Testing Cost per Hour

Currently, the die testing cost per hour ranges from $200 to $300 [18]. We assume its value in 1990 was $240 per hour and increasing at the rate of $3 per year. The rate of increase is arbitrary because it depends on where the manufacturer falls on the learning curves—data that are unavailable to the public due to the competitive nature of the business. These assumptions, when expressed mathematically, yield:

$$TCPH_t = TCPH_{1990} + 3 * (year - 1990) \ (\$/hour). \qquad (3.32)$$

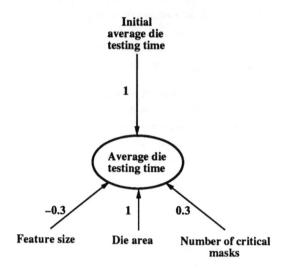

Figure 3.9: Relational diagram of the average IC die testing time.

Average Die Testing Time

The relational diagram of the average die testing time is illustrated in Figure 3.9, where the average die testing time at t ($ADTT_t$) depends upon the initial average die testing time at t_0 ($ADTT_0$), the feature size, the die area, and the number of critical masks:

1. Feature size was chosen as an average die testing time technology-driving trend because a smaller feature size increases the number of transistors on and the functionality of a chip (as seen in section 2.2), thus increasing the number of modules to test and, in the process, increasing the die testing time.

2. Die area was chosen as an average die testing time technology-driving trend because a larger die can incorporate more modules and more application specific functions, thus increasing the amount of time to test it.

3. Number of critical masks was chosen as an average die testing time technology-driving trend because it reflects the degree of manufac-

3.2. ICS SUPPLY MODEL: MICROPROCESSORS AND DRAMS

turing sophistication and the level of integration the CPU production process has reached; both factors increase the complexity of the chip and the time to test it.

The average die testing time can be expressed mathematically as follows:

$$ADTT_t = ADTT_0 * \left(\frac{FS_t}{FS_0}\right)^{-0.3} * \frac{DA_t}{DA_0} * \left(\frac{CM_t}{CM_0}\right)^{0.3} \quad (sec). \quad (3.33)$$

The average die testing time increases as the feature size decreases and as the die area and the number of critical masks increase. The values of the exponents (a_ns) in equation (3.33) were obtained by tuning the model to yield results to fall into the testing time ranges observed historically. As of 1991, the average die testing time took anywhere from 10 seconds to 1 minute [18], depending on the die's complexity. Though hard data are not available, the average die testing time in 1980 was assumed to be in the range of 15 seconds to 1.5 minutes.

All of the $|a_n|$s in equation (3.33) are ≤ 1, and each reflects an estimate of its relative importance to the behavior of the average die testing time over time. For example, a 10 percent decrease in the feature size leads to an approximate increase of 3 percent in the average die testing time [see equation (3.9)], and similarly for the other factors of equation (3.33).

IC Packaging Cost

The IC packaging cost depends on the die area, the number of pins per package, and the type of package [33]. The type of package used is dependent on the amount of power the die consumes and on the frequencies at which it is operated. For die areas less than 1.1 cm^2, the number of pins[22] usually ranges from 100 to 200, and a plastic package is recommended because the heat dissipated is within the limitations of the plastic used [33]. Even though several new plastic packages with high heat tolerance ratings are available, ceramic packages are recommended [33] for die areas in the 1.1+ cm^2 range because they can better handle the heat generated from the power dissipated in the increased die area and the higher number of package pins. Packages with up to 1,000 pins can be manufactured [101], but for the ICs within the scope of this book,

---

[22]The number of pins increases with the functionality of the die.

a package with up to 500 pins is more than adequate. For a die area less than 1.1 cm², the packaging cost is [33]:

$$PC_t = 5\ (\$)\ for\ DA \leq 1.1 cm^2 \qquad (3.34)$$

and for a die area larger than 1.1 cm², the packaging cost is [33]:

$$PC_t = 52\ (\$)\ for\ DA \geq 1.1 cm^2. \qquad (3.35)$$

We assume that the packaging costs remain constant over the study period of the model simulation.

IC Burn-in Cost

After packaging, the die goes through a burn-in phase, which costs about 25 cents [33]:

$$BIC_t = 0.25\ (\$). \qquad (3.36)$$

The burn-in cost is assumed to be constant over the model's period of study.

IC Final Test Yield

The final test yield of good dies, after burn in, is incorporated as the denominator in the IC total cost equation [equation (3.26)], making the good dies pay for the faulty ones by increasing their total cost. This final die-test yield is close to 90 percent [33] and we assume it remains constant throughout the model's period of study.

Remark

In some literature [27, 43], the historical trend of the number of transistors has been used as a technical indicator of the evolution of the semiconductor industry and its ability to manufacture high-density ICs. In this book, the chief technical indicator is feature size. In retrospect, the feature size represents the ability of the manufacturer to produce high density ICs and, in the chain of technology-driving trends, feature size drives the transistor densities to increase on a chip.

3.3 Magnetic Hard Disk Supply Model

In the early 1960s, digital magnetic recording was introduced in large computer systems because it could provide an efficient response to a data access, it was reliable, and it was low in cost [3, 58, 90]. IBM was the leading innovator in the technology [103] because it complemented its large-systems/large-database mainframe computers (which is still the bread and butter of IBM's business). As computers became smaller and the demand for them grew, the need for small, computer-resident, data storage devices emerged. Companies such as Conner Peripherals, IBM, Tandon, and others designed and produced several types of magnetic storage devices and supplemented them with portable and flexible magnetic diskettes. To differentiate the diskette from the computer-resident magnetic disk, the notion of magnetic hard[23] disk storage emerged. The data retained in magnetic storage are permanent because the magnets or the bit cells where the data are stored are permanent magnets.

The magnetic storage technology has been challenged by optical and magneto-optical storage technologies. However, with several improvements to the magnetic recording media, the read and write heads, the channel's electronic signal processing, and the packaging techniques, the magnetic technology has so far been able to maintain the lead among all the information-recording sectors of the computer industry. For more details on the technological improvements, check the following references [3, 5, 58, 90, 103, 107].

Subsection 3.3.1 presents historical data on the physical characteristics trends of magnetic hard disks; subsection 3.3.2 presents historical data on capabilities and price trends of rigid magnetic disks; subsection 3.3.3 presents the assumptions and the terminology used in the development of the supply model of magnetic hard disks; and subsections 3.3.4 through 3.3.11 present the mathematical formulation of the magnetic hard disk supply model.

---

[23]In the literature, the words "hard" and "rigid" are synonymous when used in describing a magnetic disk storage device.

3.3.1 Historical Data on the Physical Characteristics of Magnetic Hard Disks

Magnetic hard disk production is a worldwide activity, and several companies play major roles in moving a final product to market. Multiple steps are involved in the manufacturing of each hard disk subcomponent. The higher the accuracy of production is, the higher the reliability and performance of the disk.

As presented in section 2.3, a magnetic storage system can have more than one disk, with read, write, and erase heads for each disk surface. The disk surface is a series of concentric circular tracks, where each track is formed by a concatenation of magnetic bit cells. The heads are moved by an actuator, controlled by a servomechanism which computes the distance the heads must move before reaching the required data. The data are received and transmitted by the hard disk through a channel. This channel is characterized by a maximum data rate at which the hard disk can operate and a microcontroller that coordinates the data transfer.

For the purposes of reviewing historical trends in the physical characteristics of magnetic hard disks, the subcomponents of the magnetic storage devices are grouped in three categories. The first category is the disk and the magnetic medium, the second is the read/write/erase heads and the actuator, and the third is the data channel. Figure 3.10[24] presents the technology-driving trends [3] of the first and second categories in a log-linear graph, where the vertical axis indicates the base 10 logarithm of their values in nanometers and the horizontal axis indicates the year of the IBM's DASD[25] market introduction. The IBM DASDs product numbers are listed directly above the horizontal axis. For instance, the products IBM 3380 and 3380E were introduced in 1982 and 1985, respectively.

The trends of each of the three categories of magnetic storage subcomponents listed above are discussed in the following paragraphs, as well as the behavior over time of the head-actuator setup width.

---

[24]C.H. Bajorek, "Trends in Recording and Control and Evolution of Subsystem Architectures for Data Storage," **IEEE Proceedings, COMPEURO'89: VLSI and Computer Peripherals**, ©1989 IEEE. Reprinted with permission of IEEE Press.

[25]DASD is equivalent to Direct Access Storage Device.

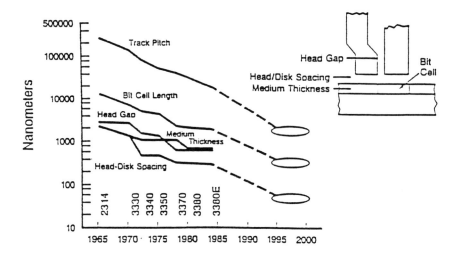

Figure 3.10: DASD recording system scaling. Source: [3]. ©1989 IEEE.

Disk and Magnetic Medium

The disk and magnetic medium technology-driving trends are the disk track pitch, the bit cell length, and the medium thickness. Keep in mind that as the track pitch decreases, the areal density of the disk increases and, as the bit cell length decreases, both the areal and the linear densities increase. The medium thickness measures the thickness of the magnet-sensitive film coating on the disk. Figure 3.10 shows that it exhibits a decreasing trend. This is due to the use of thinner films of magnetic materials in the production process. A thinner media has also improved the track pitch and the bit cells layout and increased the areal density of the disk. The trends of the track pitch, the bit cell length, and the medium thickness in Figure 3.10 can be expressed mathematically as follows:

$$TP_t = 310 * 10^{-0.066*(t-1965)} \ (micron) \qquad (3.37)$$
$$BCL_t = 13.5 * 10^{-0.049*(t-1965)} \ (micron) \qquad (3.38)$$
$$MT_t = 2.11 * 10^{-0.0295*(t-1965)} \ (micron) \qquad (3.39)$$

where TP_t represents the track pitch in microns, BCL_t represents the magnetic bit cell length in microns, MT_t represents the medium thickness in microns, and t represents the year in which the track pitch, the bit cell length, and the medium thickness are calculated.

Read/Write/Erase Heads and Actuator

The inductive read/write/erase head technology-driving trends are measured by the head gap spacing and the head-medium spacing. Illustrated in the top right of Figure 3.10, the head gap is shown to be the distance separating the poles of the magnetic core of the head, and the head-medium spacing is shown to be the distance at which the head flies above the disk surface. A smaller head gap complements a smaller bit cell length, improving both the linear and areal densities of the disk. A smaller head-medium spacing decreases the data misregistration errors, complements the rotational speed of the disk, and improves the data rates (megabytes per second) at which the disk can operate. The trends of the head gap spacing and the head-medium spacing are traced in Figure 3.10. They can be approximated mathematically by equations (3.40) and (3.41), respectively:

$$HGS_t = 2.8 * 10^{-0.04*(t-1965)} \ (micron) \qquad (3.40)$$

$$HMS_t = 2.11 * 10^{-0.055*(t-1965)} \quad (micron) \qquad (3.41)$$

where HGS_t represents the head gap spacing in microns, HMS_t represents the head-medium spacing in microns, and t represents the year in which the head gap and the head-medium spacings are calculated.

No trends were available on actual performance measures of the servo and the actuator[26], such as data-access times, seek times, and latency. Nevertheless, improvements have been made to integrate the head, the actuator, and the servo in one device by using a technique called micromechatronics [90], which offers improvements in the overall performance levels of the disk. The effect of this integration on the hard disk's attributes will be reflected through the cost per megabyte trend, to be presented later in the model formulation subsection. What follows is a description of the head-actuator setup factor and how it affects the magnetic hard disk's volumetric density.

Head-Actuator Setup

As the technology-driving trends in Figure 3.10 improve over time, the areal density and the data rate of the magnetic hard disk improve. Furthermore, the volumetric density of the disk is improved by the reduced head-medium spacing, the reduced medium thickness, and the overall smaller head-actuator setup [90]. A decrease in the head-actuator setup width increases the number of disks to be packaged in a fixed height storage system. As of 1991, the head-actuator setup width was in the 0.4 to 0.5 centimeter range. Unfortunately, no head-actuator setup width data for other years were available, so no trend could be estimated from historical data. We assumed the head-actuator setup width to be decreasing at approximately the same rate as the technology-driving trends provided in Figure 3.10, making it possible to express its behavior over time as follows:

$$HASW_t = HASW_0 * 10^{-0.04*(t-1991)} = 0.45 * 10^{-0.04*(t-1991)} \quad (cm)$$
$$(3.42)$$

where $HASW_t$ represents the head-actuator setup width in centimeters, $HASW_0$ represents the initial value of the head-actuator setup width—assumed to be 0.45 cm as of 1991—and 0.04 represents an average of

---

[26]The servo and actuator performance measures depend on the configuration and the physical dimensions of the magnetic hard disk—all manufacturer dependent characteristics on which no trends were available.

the slopes of the lines in Figure 3.10. The t term represents the year in which the head-actuator setup width is calculated.

Data Channel

The hard disk's data channel performance is characterized by its data bandwidth, or data rate, measured in megabytes per second. A microcontroller is located at the end of the channel, decoding and queuing disk data-access commands. As the disk responds with the requested data, the microcontroller receives the data and feeds them to the channel, thereby controlling the data rate. The data rate of the channel must be at least equal to the data response rate of the disks. The technology-driving trends of the disk microcontroller are the same as those for microprocessors, so the MIPS rating of the microprocessors is used as a *relative* measure to reflect the performance of the data channel. Since most microcontrollers have been designed along the lines of the CISC architecture [72], the CISC MIPS rating behavior of the ICs supply model will be used in the equation capturing the effects of the data channel on the overall supply of the magnetic hard disk.

3.3.2 Historical Data on Magnetic Hard Disk Capabilities and Price Trends

The following paragraphs present the price per megabyte and areal density trends of the magnetic hard disk storage. The price per megabyte encompasses the price contribution of all the hard disk's components, and the areal density encompasses the linear and track densities of the hard disk. The volumetric density of the disk depends on the head-actuator setup width, the trend of which, unfortunately, was not available to us[27] so that no historical trend on volumetric density could be estimated.

Magnetic Hard Disk Price per Megabyte Trend

Two log-linear graphs are presented in Figure 3.11[28], the first being the price per megabyte of large-capacity magnetic hard disks and the second

---

[27]In equation (3.42), the behavior of the head-actuator setup width over time was assumed.

[28]H. Takata, "Future Trend of Storage Systems," **IEEE Proceedings, COMPEURO'89: VLSI and Computer Peripherals**, ©1989 IEEE. Reprinted with permission of IEEE Press.

3.3. MAGNETIC HARD DISK SUPPLY MODEL

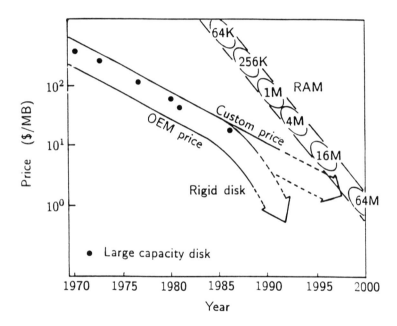

Figure 3.11: Actual price per megabyte of magnetic hard disk storage versus time. Source: [90]. ©1989 IEEE.

the price per megabyte of DRAM chips [90]. The vertical axis indicates the base 10 logarithm of the prices in $/MB and the horizontal axis indicates the time. The large capacity rigid disk part of Figure 3.11 has two lines, one indicating the custom price trend and another indicating the original equipment manufacturer (OEM) price trend. The two lines are parallel, and their values differ by almost 100 percent. The behavior of the actual price per megabyte trend of large capacity magnetic hard disks can be expressed mathematically as follows:

$$PMBD_t = 404.57 * 10^{-0.0857*(t-1970)} \quad (\$/MB) \qquad (3.43)$$

where $PMBD_t$ represents the actual dollar price per megabyte, and t represents the year in which the actual price per megabyte is calculated.

If we assume that the manufacturer marks up the direct costs of the large capacity magnetic hard disks by 200 percent before selling them to the retail companies, the cost per megabyte can be computed from equation (3.43) by dividing the price by 3:

$$CMBD_t = 134.86 * 10^{-0.0857*(t-1970)} \quad (\$/MB). \qquad (3.44)$$

Equation (3.44) is only a cost per megabyte estimation to be used in tuning the results of the magnetic storage model. The manufacturer's actual costs could be lower or higher, depending on the manufacturer's position on the learning curves and the attributes of the manufactured hard disks.

Hard Disk Areal Density Trend

The areal density trends of several magnetic storage devices are presented in Figure 3.12[29] as log-linear graphs, where the vertical axis indicates the magnetic device density in bits per mm^2, and the horizontal axis indicates the product introduction year. Each device trend is labeled. For instance, the rigid disk's density trend is labeled RD, the flexible disk's FD, the home video tape's HV, the professional video tape's PV, the audio tape's AT, and the data tape's DT [58]. The rigid disk's areal density is the trend of interest and can be expressed mathematically as follows:

$$RDAD_t = 64.53 * 10^{0.138*(t-1965)} \quad (bits/mm^2) \qquad (3.45)$$

---

[29] C.D. Mee and E.D. Daniel, **Magnetic Recording Handbook**, ©1989 McGraw-Hill. Reprinted with permission of McGraw-Hill Book Company.

3.3. MAGNETIC HARD DISK SUPPLY MODEL

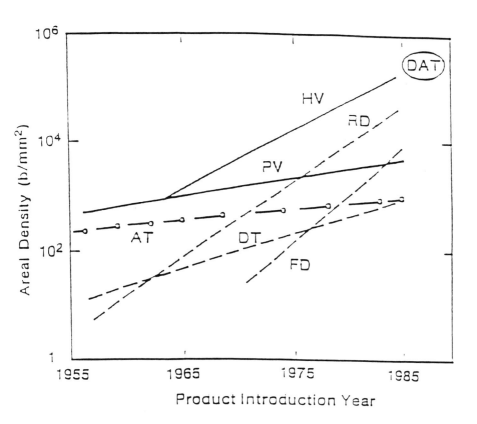

Figure 3.12: Areal densities of magnetic storage devices in bits per mm^2 versus time. Source: [58]. ©1989 McGraw-Hill.

where $RDAD_t$ represents the rigid disk's areal density in bits per mm², and t represents the year in which the areal density is calculated.

3.3.3 Model Assumptions and Terminology

The digital file storage supply model incorporates equations capturing the behaviors of magnetic hard disk attributes, specifically, the cost per megabyte, the data rate, and the areal density of rigid magnetic storage media. Before presenting the mathematical formulation of the model, several assumptions are listed. This is followed by a list defining the variables and the parameters used in the model.

Model Assumptions

In modeling the dynamics of the supply of magnetic hard disks, the following assumptions are made:

1. Each disk is magnetically coated on both surfaces.

2. Each disk surface has one single read/write/erase head.

3. All the disks surfaces can be read or written at the same time.

4. When the storage system is powered on, the disks reach a design specific rotational speed and maintain this speed throughout the system's operations.

5. The storage system's microcontroller and channel have a constant operational speed and a constant bandwidth, respectively. Consequently, the system has a constant data rate.

6. The storage system dimensional specifications, such as the height and the disk diameter, are assumed to remain constant over the period of study.

7. Technological trends of the past in overcoming physical manufacturing barriers continue during the period of study.

8. Past trends in magnetic hard disk manufacturing yields will continue to increase or, at worst, remain constant.

9. Past trends in magnetic hard disk testing will continue to improve to deal with the higher density disks, thereby improving their reliability.

3.3. MAGNETIC HARD DISK SUPPLY MODEL

Definitions and Terminology

In presenting the model's equations, several parameters and variables are introduced, and they are defined as follows:

0	time index for the first year of the study period
t	time index
TP	track pitch, in microns
BCL	bit cell length, in microns
HGS	head gap spacing, in microns
HMS	head-medium spacing, in microns
MT	magnetic medium thickness, in microns
DT	metal disk thickness, in microns
$HASW$	head-actuator setup width, in cm
$RSUP$	hard disk support area radius, in cm
$RSTO$	hard disk storage area radius, in cm
DD	hard disk diameter, in cm
DH	height of the hard disk box, in cm
$DPER$	number of disks per centimeter of box height
X	number of tracks on each disk surface
$RDAD$	rigid disk areal density, in megabytes per cm^2
TC	track capacity, in megabytes
LD	linear density per track, in megabytes per cm
TD	track density across the disk surface, in tracks per cm
AC	areal capacity per disk surface, in megabytes
AD	areal density per disk surface, in megabytes per cm^2
VC	volumetric capacity of the storage system, in megabytes
VD	volumetric density of the storage system, in megabytes per cm^3
$MIPS$	number of million instructions per second
$DRPM$	number of disk rotations per minute
DR	storage system data rate, in megabytes per second
CMB	cost per megabyte (model results), in dollars per megabyte
$CMBD$	cost per megabyte (actual data), in dollars per megabyte
TCD	total cost of the magnetic hard disk storage system, in dollars

The mathematical formulation of the magnetic hard disk model will proceed as follows: in subsection 3.3.4 the radius of the magnetic storage media is computed; in subsection 3.3.5 the number of disk recording tracks is calculated; in subsection 3.3.6 the track capacity and density are computed; in subsection 3.3.7 the data rate computation is presented, followed by a calculation of the optimal magnetic storage radius when the data rate is constant; in subsections 3.3.8 and 3.3.9 the areal and volumetric capacities and densities are calculated, respectively; in subsection 3.3.10 the hard disk cost per megabyte is expressed as a function of several magnetic and semiconductor technology-driving trends; and finally, in subsection 3.3.11 the total hard disk cost is computed.

3.3.4 Magnetic Storage Radius

As shown earlier, a magnetic hard disk consists of a stack of disks, each separated from the other by a head-actuator setup width. The disks are coated on each surface with magnetic media capable of permanently storing the digital data in a format of polarized magnets. Each disk surface is divided into X number of tracks, separated from each other by $X - 1$ number of track pitches. However, part of the disk surface is used as a rotational support, so the width of the magnetic storage surface, or what we refer to as the magnetic storage radius, is equal to:

$$RSTO = \frac{DD}{2} - RSUP \quad (cm) \qquad (3.46)$$

where $\frac{DD}{2}$ is equivalent to the disk radius and $RSUP$ is the disk support radius. See Figure 3.13 for an illustration of the disk parameters used in equation (3.46).

3.3.5 Number of Disk Recording Tracks

We assume that the width of a track is equal to one bit cell length, and relating the number of tracks, the number of track pitches, and the storage radius, the following equation is obtained:

$$BCL_t * X + TP_t * (X - 1) = RSTO * 10^4. \qquad (3.47)$$

The storage radius ($RSTO$) is actually the distance between the support radius ($RSUP$) and the rim of the disk. We refer to it as a storage "radius" in order to be consistent with the fact that this distance is filled with circular tracks. From equation (3.47), the number of tracks X is computed as:

$$X = \frac{RSTO * 10^4 + TP_t}{BCL_t + TP_t} \approx \frac{RSTO * 10^4}{TP_t} \qquad (3.48)$$

3.3. MAGNETIC HARD DISK SUPPLY MODEL

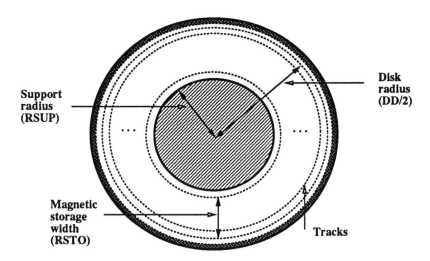

Figure 3.13: Illustration of the dimensional parameters of a magnetic hard disk.

where X is approximated as the ratio of the storage radius and the track pitch at t because the value of the bit cell length is negligible when compared to that of the track pitch.

3.3.6 Track Capacity and Density

Each track has the same number of sectors, and, since the disk is rotating at a constant speed and has a constant data rate, each sector—and, consequently, each track—has the same number of bits [58]. Since the support radius ($RSUP$) has the smallest track in circumference and all the tracks have the same capacity, the number of bits on this track sets the standard number of bits that each track can have. The track capacity in megabytes[30] is:

$$TC_t = \frac{2 * \pi * RSUP * 10^4}{BCL_t} \ (bits) \ = \ \frac{2 * \pi * RSUP * 10^4}{BCL_t * (8 * 10^6)} \ (MB). \tag{3.49}$$

---

[30] Since there are 8 bits to a byte, 8 * 10^6 is the product used to transform the track capacity values from bits to megabytes.

The linear bit density per track is equal to the number of bit cells in one centimeter, and its equation is:

$$LD_t = \frac{10^4}{BCL_t * (8 * 10^6)} \quad (MB/cm) \tag{3.50}$$

and the track density across the storage surface is:

$$TD_t = \frac{10^4}{TP_t} \quad (tracks/cm). \tag{3.51}$$

3.3.7 Hard Disk Data Rate

Before computing the areal and volumetric capacities of the storage system, since the data rate has been mentioned several times, it is appropriate at this point to present its formulation. What follows is the mathematical expression of the data rate, with a description of each of the expression's factors. The data rate equation is:

$$DR_t = TC_t * \frac{DRPM_t}{60} * \lfloor (DPER_t * DH) \rfloor \quad (MB/sec). \tag{3.52}$$

The first term in the data rate equation is the *track capacity (TC)* in megabytes, the formulation of which is given by equation (3.49). The second term is the *number of disk rotations per minute (DRPM)*, expressed in seconds by dividing it by 60. The DRPM depends on the rotating motor speed, on the channel's bandwidth, and on the head and media characteristics handling the read and write accesses. For the model, the DRPM trend was assumed to increase at 100 rotations per year since 1980, with an initial value of 3,000. As of 1991, the DRPM was in the 3,600–4,200 range and, based on conversations with colleagues, the number of disk rotations per minute was expected to reach 5,000 by the year 2000. The DRPM equation over time is:

$$DRPM_t = 3,000 + (t - 1980) * 100 \quad (rot/min) \tag{3.53}$$

where t represents the year in which the DRPM is calculated. The DRPM will saturate over time due to the high heat generated from the rotations and the high power it absorbs to reach them [58]. In 1991, an average value of DRPM is 4,100 rotations per minute. (Most manufacturers now rely on techniques other than increased rotational speed to improve their data rates.)

3.3. MAGNETIC HARD DISK SUPPLY MODEL

The third term in equation (3.52) captures the *number of disk surfaces* off which data can be read or onto which data can be written. For a given storage system box height, the number of disks that can fit depends on the head-actuator setup width, the disk thickness, the medium thickness, and the head-medium spacing. Since each disk surface has one read/write/erase head and since all the surfaces can be read or written at the same time, the actual data rate of the system is equal to the data rate per one disk surface, which is equal to the product of the first and the second terms of equation (3.52), multiplied by the number of surfaces written or read. Because the disks are coated on each side, each disk has two heads associated with it, two head-actuator setups, two media surfaces, two head-medium spacings, and one disk thickness; the equation for the number of disks per centimeter is:

$$DPER_t = \frac{1}{[2*(MT_t+HMS_t)+DT_t]*10^{-4}+2*HASW_t} \quad (disks/cm) \quad (3.54)$$

$$\approx \frac{1}{2*HASW_t} \quad (disks/cm). \quad (3.55)$$

Historically, the number of disks per centimeter ($DPER$) was strongly influenced by improvements to the head-actuator setup width ($HASW$), the assumed trend of which was given as equation (3.42). In equation (3.52), the product of the storage system's height (DH), in centimeters, and the number of disks per centimeter ($DPER$) is equal to the total number of disks assembled in the storage system.

The data rate of a storage system is not an attribute that could easily have a general behavioral trend over time. This is due to the data rate's dependency on the configuration and physical dimensions of the storage system, both of which do not follow a standard dimension scheme. Nevertheless, the data rate trend is on the increase.

Optimal Hard Disk Storage Radius

For a storage system with a constant data rate, it is demonstrated in [58] that the track capacity and, consequently, the storage capacity are maximized by having the support radius, $RSUP$, equal to half the disk radius:

$$RSUP_{opt} = \frac{DD}{4} \quad (cm). \quad (3.56)$$

From equations (3.46) and (3.56), the optimal width of the storage surface, or what we refer to as the optimal storage radius, is equal to the support radius:

$$RSTO_{opt} = RSUP_{opt} = \frac{DD}{4} \quad (cm). \tag{3.57}$$

3.3.8 Areal Capacity and Density

As shown earlier, all the tracks on a disk surface have the same capacity. So the areal capacity per disk surface is simply equal to the product of the track capacity and the number of tracks:

$$AC_t = TC_t * X \quad (MB). \tag{3.58}$$

From equations (3.50) and (3.51), the areal density can be computed as the product of the linear bit density and the track density:

$$AD_t = LD_t * TD_t \quad (MB/cm^2). \tag{3.59}$$

3.3.9 Volumetric Capacity and Density

The volumetric capacity of the storage system is equal to the number of disks in the system multiplied by twice the areal capacity per disk surface, since the disks are coated on both surfaces:

$$VC_t = 2 * AC_t * \lfloor (DPER_t * DH) \rfloor \quad (MB) \tag{3.60}$$

and the volumetric density is equal to the product of twice the disk areal density and the number of disks per centimeter:

$$VD_t = 2 * AD_t * \lfloor DPER_t \rfloor \quad (MB/cm^3). \tag{3.61}$$

3.3.10 Magnetic Hard Disk Cost per Megabyte

The final equations of the model compute the cost per megabyte of magnetic hard disk storage and, ultimately, the total cost of the storage system. As discussed earlier, the storage system has three groupings of subcomponents, and each of these subcomponents has a set of technology-driving trends. Figure 3.14 illustrates, in a relational diagram, the effect of each subcomponent's set of technology-driving trends on the overall cost per megabyte of magnetic storage:

3.3. MAGNETIC HARD DISK SUPPLY MODEL

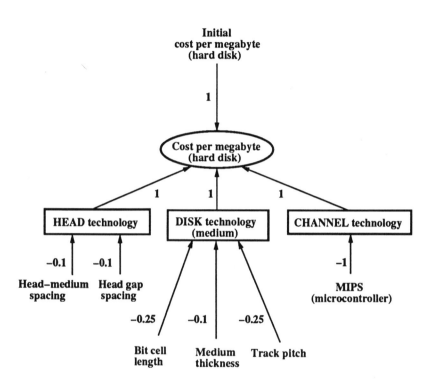

Figure 3.14: Relational diagram of the cost per megabyte of a magnetic hard disk.

1. Head gap spacing and head-medium spacing were chosen as magnetic hard disk head technology-driving trends and, indirectly, magnetic hard disk cost-driving trends, because achieving lower spacings requires much scientific research, effort, and R&D money, all of which increase the magnetic hard disk cost per megabyte.

2. Bit cell length, medium thickness, and track pitch were chosen as disk technology-driving trends and, indirectly, magnetic hard disk cost-driving trends, because a smaller bit cell length is harder to deposit on the disk surface and needs fancier head and servomechanism designs to handle the higher bit densities resulting from a smaller magnetic cell length; similarly, it is scientifically challenging to achieve lower track pitches and lower medium thicknesses, all costly R&D outlays for disk production, and factors that increase the magnetic hard disk cost per megabyte.

3. MIPS was chosen as a channel technology-driving trend and, indirectly, a magnetic hard disk cost-driving trend, because as the MIPS rating of the channel's microcontroller improves, the price per MIPS decreases (see actual data in section 3.2 and the ICs supply model results in section 4.3). Thus, if the data rate of the disk is assumed to remain the same, the cost of the channel decreases and, consequently, the cost per megabyte of magnetic hard disk.

Expressed mathematically, the cost per megabyte is equal to:

$$CMB_t = CMB_0 * HEAD_t * DISK_t * CHANNEL_t \quad (\$/MB) \quad (3.62)$$

where

$$HEAD_t = \left(\frac{HGS_t}{HGS_0}\right)^{-0.1} * \left(\frac{HMS_t}{HMS_0}\right)^{-0.1} \quad (3.63)$$

$$DISK_t = \left(\frac{BCL_t}{BCL_0}\right)^{-0.25} * \left(\frac{MT_t}{MT_0}\right)^{-0.1} * \left(\frac{TP_t}{TP_0}\right)^{-0.25} \quad (3.64)$$

$$CHANNEL_t = \left(\frac{MIPS_t}{MIPS_0}\right)^{-1} \quad (3.65)$$

and the initial cost per megabyte, CMB_0, is computed from equation (3.44) for the initial year in the period of study. (It is assumed that CMB_0 includes the cost of labor and raw materials, in particular the actuator and the servomechanism costs mentioned earlier.) Equations

(3.63) and (3.64) reflect how the cost per megabyte of magnetic hard disk storage in equation (3.62) increases as the head gap spacing, the head-medium spacing, the bit cell length, the medium thickness, and the track pitch decrease. Equation (3.65) reflects how an increase in the MIPS rating of the storage system's microcontroller decreases the actual cost per MIPS of the chip and, consequently, the cost per megabyte of magnetic storage. The values of the exponents, (a_ns), in equations (3.63), (3.64), and (3.65) were obtained by tuning the model to yield results to match the actual cost per megabyte data of hard disk storage provided in subsection 3.3.2. All of the $|a_n|$s are ≤ 1, and each reflects an estimate of its relative importance to the behavior of the cost per megabyte over time. For example, a 10 percent decrease in the bit cell length leads to an approximate increase of 2.5 percent in the cost per megabyte [see equation (3.9)], and similarly for other factors of equations (3.63), (3.64), and (3.65).

3.3.11 Magnetic Hard Disk Total Cost

Finally, the total cost of the magnetic hard disk storage system is equal to the product of the volumetric capacity and the cost per megabyte:

$$TCD_t = VC_t * CMB_t \quad (\$). \qquad (3.66)$$

3.4 Color CRT Display Supply Model

The CRT is still, after 100 years of experience in its manufacturing, a large, heavy, and power-hungry display technology. Its popularity can be attributed to its diversified range of applications, its fast response time, and its low cost relative to the other display technologies [92]. As of 1991, the CRT has the highest market share of displays sold [93], and users continue to appeal to the monitors' manufacturers to reduce their prices, increase the CRT's resolution, and eliminate the glare [2]. Other than in home televisions, the CRT capabilities are most apparent in the computer workstation displays, where CAD and computer simulations and animations are the most prevalent applications and where the CRT's speed of response in an interactive environment is key.

The size and power consumption of the CRT are a consequence of the physics of its design. These disadvantages have been exploited by leaner battery-operated displays, which may overtake the CRT's market share lead, at least in the computer market. The liquid crystal based display (LCD) is the leading challenger and has the highest potential of displacing the CRT's market share. Already 10-inch color LCDs are available in portable computers [70], and before long workstation LCDs will be developed and sold as independently packaged computer displays, as CRTs are now.

Most workstation CRT displays measure 19 to 20 inches on the diagonal, and the screens are available in color or B&W. The CRT supply model, presented later in this section, captures the cost per megapixel trend of a 19-inch color CRT and relates it to the resolution technology-driving trends such as the metal shadow mask hole pitch, the speed of the screen-driving circuitry, and the enhancements to the electron beams generation, acceleration, and deflection hardware used. Before presenting the mathematical formulation of the color CRT display supply model, historical data on the color CRT's physical characteristics are presented in subsection 3.4.1, and historical data on the color CRT's capabilities and price trends are presented in subsection 3.4.2.

3.4.1 Historical Data on the Physical Characteristics of Color CRT Displays

As discussed in section 2.4, the sharpness and the resolution of a CRT picture are inversely proportional to the size of the pixel or the size of the smallest picture element that can be displayed within the limitations and parameters of the display structure and hardware. In workstations today, 1+ megapixel color CRT displays are very common. Display hardware drivers can be expensive at times, depending on the application and the color palettes desired [33].

In response to the increased demand for higher resolutions and more colors per pixel, CRT production has been subjected to several enhancements and design changes over the years. Recent VLSI improvements have shifted the focus for improved resolution from the CRT driving circuitry to the electron beams generation, focus, and deflection hardware, and to the manufacture of the tube, including the metal shadow mask used with color CRTs [8, 14].

When excluding the driving circuitry, the main parts of the color CRT structure are the tube, the electron guns, the electron beams' accelerating, focusing, and deflecting apertures, and the metal shadow mask.

Since the 19-inch color CRT workstation monitor was only introduced in 1985 [16, 33], it has a brief history. The current leading manufacturers are Japanese, companies like Hitachi, Mitsubishi, NEC, and Sony, whose engineering facilities are spread between the far eastern rim of Asia and the western United States, so obtaining any manufacturing trends has been very difficult. Several companies, like Stanford Resources and Tannas Electronics, have compiled reports and books on the CRT production trends, but they are costly and largely limited to future projections based on past behavior, rather than historical data of technology-driving trends[31]. The only actual physical characteristic data found track the hole pitch trend of the metal shadow mask.

Metal Shadow Mask Hole Pitch

The metal shadow mask hole pitch data are listed in column 5 of Table 3.7, with values dating back to 1982. If one traces a curve to the hole pitch's trend over time, the result can be expressed mathematically as follows:

$$HP_t = 0.6 * 10^{-0.04*(t-1982)} \quad (mm) \tag{3.67}$$

where HP_t represents the hole pitch in millimeters, and t represents the year in which the hole pitch is calculated. The number of holes per inch (HPI_t), or the maximum number of pixels per inch (MaxPPI_t), is equal to:

$$HPI_t = MaxPPI_t = \frac{25.4}{HP_t} \quad (holes/inch \ \ or \ \ Max \ \#pixels/inch)$$
$$\tag{3.68}$$

where 25.4 is equal to the number of millimeters per inch.

It was stated earlier that improved ICs manufacturing capabilities will lead to faster and denser microcontrollers and DRAMs, respectively, and, therefore, increase the number of holes per inch. Another important technology that enables a continuous increase in the number of holes per inch—and, consequently, the CRT's resolution—is lithography. Lithography is used to deposit the color phosphors on the inner face plate of the bulb, one color dot at a time. The more sophisticated lithography is, the more in pace the phosphors resolution on the face plate will be with the holes resolution of the metal shadow mask. Unfortunately, very small hole pitches reduce the metal shadow mask production yields and,

---

[31] This is the case with the reports compiled by Stanford Resources.

Table 3.7: Actual market data of 19- and 20-inch color CRT displays. Sources: [1, 2, 16, 28, 39, 60, 76, 82, 84, 95].

1	2	3	4	5	6	7	8
			Color CRTs				
Year	Name Brand	Diagonal Size (inch)	Resolution (HxV) or (MP)	Hole Pitch (mm)	Refresh Rate (Hz)	Bandwidth (MHz)	Price (Actual Data) ($)
1982	n.a.	n.a.	n.a.	0.6	n.a.	n.a.	n.a.
1985	Sony	19	1	n.a.	n.a.	n.a.	8K
1986	Sony	19	1	n.a.	n.a.	n.a.	6K
1987	Mitsubishi	20	1280x1024	0.31	50-75	100	3.8K
1988	n.a.	19	1024x1024	n.a.	n.a.	n.a.	2.5K-4K
1989	Sony	19	1	n.a.	n.a.	n.a.	5K
1989	Sony	19	2048x1536	0.19-0.2	80	262	n.a.
1989	NEC	19	1024x768	0.31	56-80	65	3.2K
1989	n.a.	n.a.	1024x768	n.a.	n.a.	n.a.	3K
1989	n.a.	n.a.	1600x1280	n.a.	n.a.	n.a.	5K-10K
1991	Sony	19	1280x1024	0.28	n.a.	n.a.	3K

consequently, increase the display costs, thus making the monitors prohibitively expensive for general workstation users.

3.4.2 Historical Data on Color CRT Display Capabilities and Price Trends

Capabilities and price data of the 19- and 20-inch color computer monitors are presented in Table 3.7. (Unavailable data are listed as n.a.) Under column 2, the CRT manufacturers are listed. Sony is among the listed manufacturers; however, Sony uses a different shadow mask technology than that considered in the CRT model presented in this section. Still, Sony's monitor resolutions and prices reflect the general trends of the industry. Their color CRTs use the Trinitron specifications: each of the three color phosphors—red, green, or blue—are deposited in vertical stripes along the inner screen of the tube, and each stripe has a different color than those adjacent to it. The deposition pattern is repeated along the screen, and three electron guns, one per color, are used to excite the phosphors, simultaneously, while scanning the screen and generating the image. The shadow mask could be considered as a striped metal sheet, located behind the screen, where each stripe corresponds to a particular color phosphor. The mask technology considered in this section's CRT supply model uses a perforated metal shadow mask, where each perforated hole corresponds to three phosphor dots, each with a different color. The three electron beams must go through the mask hole and hit the dots, simultaneously, for the corresponding color light to be emitted.

Most of the monitors listed on Table 3.7 measure 19 inches on the diagonal. Their *resolutions* are listed in column 4 and all exhibit the 4:3 aspect ratio, except for one, which can be verified by examining the number of horizontal pixels times the number of vertical pixels (HxV) displayed. Note that the *number of pixels* is either smaller or equal to the total number of perforated holes in the mask. The total number of pixels depends on the screen-driving circuitry and on the ability of the apertures to focus and deflect the beams correctly at very high bandwidths.

The *refresh rates* of the CRTs are listed in column 6 and range from 50 to 80 image scans per second, mostly noninterlaced. Their values are mainly flicker dependent [92].

In column 7, the *bandwidths* of the CRTs reflect how fast the electron guns must switch per second to scan and refresh the screen to avoid flicker. The values are a function of the resolution and the refresh rates used in the CRT design.

Finally, the *price* data of the monitors are listed in column 8 of Table 3.7. These are single unit prices, and bulk[32] prices are often only half as much [16].

Price per Megapixel

A price per megapixel trend is provided in column 2 of Table 3.8. It is computed by dividing the prices of the monitors in Table 3.7 by their corresponding resolutions. If an exponential curve is fitted to the data to capture the behavior of the price per megapixel over time, it can be expressed as follows:

$$PMPD_t = 8,000 * 10^{-0.085*(t-1985)} \quad (\$/MP) \qquad (3.69)$$

where $PMPD_t$ represents the actual dollar price per megapixel, and t represents the year in which the actual price per megapixel is calculated. Assuming a 200 percent cost markup was used by the manufacturers [33] in setting the prices in Table 3.7, the cost per megapixel trend can be obtained by dividing each price per megapixel by 3:

$$CMPD_t = 2,700 * 10^{-0.085*(t-1985)} \quad (\$/MP). \qquad (3.70)$$

Number of Pixels per Inch

The number of pixels per inch (PPID) values of the CRTs in Table 3.7 are listed in column 3 of Table 3.8. $PPID_t$ can be computed from the following resolution equation:

$$RES_t = \frac{(HS_{in} * PPID_t) * (VS_{in} * PPID_t)}{10^6} \quad (megapixels) \qquad (3.71)$$

where RES_t represents the display's resolution in megapixels. When factored out of equation (3.71), $PPID_t$ can be expressed as:

$$PPID_t = \sqrt[2]{\frac{RES_t * 10^6}{HS_{in} * VS_{in}}} \quad (pixels/inch) \qquad (3.72)$$

---

[32]Bulk is equivalent to 10,000+ monitors.

3.4. COLOR CRT DISPLAY SUPPLY MODEL

Table 3.8: Actual price per megapixel and number of pixels per inch market data of 19- and 20-inch color CRT displays. Sources: [1, 2, 16, 28, 39, 60, 76, 82, 84, 95].

Color CRTs		
1	2	3
Year	Price per Megapixel (Actual Data) ($)	#Pixels per inch
1985	8,000	76
1986	6,000	76
1987	2,900	84
1988	2,380-3,800	67
1989	2,400-5,000	76-135
1991	2,300	84

where the horizontal size (HS) and the vertical size (VS) of the CRT can be computed from the diagonal size (DS) by using the 4:3 aspect ratio:

$$HS_{in} = \frac{4}{5} * DS_{in} \quad (inch) \tag{3.73}$$

$$VS_{in} = \frac{3}{4} * HS_{in} = \frac{3}{5} * DS_{in} \quad (inch). \tag{3.74}$$

(The 10^6 factor is used to convert the resolution values from pixel to megapixel, and vice versa.)

3.4.3 Model Assumptions and Terminology

The supply model for computer displays incorporates equations capturing the attributes of 19-inch color CRT monitors, in particular, the cost per megapixel, the bandwidth, and the number of pixels per inch and its effect on the display's resolution. Before presenting the mathematical formulation of the model, several assumptions are listed. This is followed by a list defining the variables and the parameters used in the model.

Model Assumptions

In modeling the dynamics of the supply of color 19-inch CRT displays, the following assumptions are used:

1. The display's dimensional specifications, like the diagonal size and weight, are assumed to remain constant over the period of study. The diagonal size is measured in inches.

2. Technological trends of the past in overcoming physical manufacturing barriers continue during the period of study.

3. Past trends in color CRT display manufacturing yields will continue to increase or, at worst, remain constant.

4. Past trends in color CRT display testing will continue to improve to deal with the higher resolution monitors and the higher power consumption associated with them, thereby improving the reliability of the CRTs.

Definitions and Terminology

In presenting the model's equations, several parameters and variables are introduced, and they are defined as follows:

0	time index for the first year of the study period
t	time index
DS	diagonal size of the CRT, in inches
HS	horizontal size of the CRT, in inches
VS	vertical size of the CRT, in inches
HP	hole pitch of the metal shadow mask, in millimeters
HPI	number of holes per inch in the metal shadow mask
$PPID$	number of pixels per inch (actual data)
PPI	number of pixels per inch (model results)
$MaxPPI$	maximum number of pixels per inch
RES	resolution of the display, in megapixels
$MaxRES$	maximum resolution of the display, in megapixels
$MDPUL$	number of metal shadow mask defects per unit length
MY	metal shadow mask production yield, in percent
RR	screen refresh rate, in Hertz
DB	display bandwidth, in megaHertz
$MIPS$	number of million instructions per second
$MEMORY$	capacity of a DRAM, in megabytes
$PMPD$	price per megapixel (actual data), in dollars per megapixel
$CMPD$	cost per megapixel (actual data), in dollars per megapixel
CMP	cost per megapixel (model results), in dollars per megapixel

3.4. COLOR CRT DISPLAY SUPPLY MODEL

TCM   total cost of the CRT monitor, in dollars

The mathematical formulation of the color CRT display supply model will proceed as follows: first, in subsection 3.4.4 the bandwidth of the CRT is computed; second, in subsection 3.4.5 the resolution's mathematical expression and technology-driving trends are discussed; third, in subsection 3.4.6 the metal shadow mask production yield is calculated; and fourth, in subsection 3.4.7 the cost of the CRT and the cost per megapixel are computed, including descriptions of the technology-driving trends affecting their behavior over time.

3.4.4 Bandwidth

Bandwidth has been a key factor in the CRT's response time and picture stability. It is a function of the resolution and the refresh rate of the monitor, factors depending upon the pixel density on the screen. Bandwidth is measured in megaHertz (MHz), and it is equivalent to the electron beams' switching frequency while the screen image is scanned and refreshed. The display bandwidth (DB) equation can be expressed mathematically as follows:

$$DB_t = RES_t * RR \ (MHz) \tag{3.75}$$

where the refresh rate (RR) is set to values so as to avoid flicker as the number of pixels increase and the colors become more defined [92].

3.4.5 Resolution

The display's resolution, given in equation (3.71), has almost doubled since the introduction of the first color 19-inch CRT monitor in 1985 [16, 77]. Resolution is influenced by several technology-driving trends, the most important of which are the hole pitch of the metal shadow mask and the technology-driving trends of the screen hardware drivers and the electron beams' accelerating, focusing, and deflecting apertures .

Resolution and Metal Shadow Mask Hole Pitch

The smaller the hole pitch, the higher the resolution. For a particular hole pitch, the maximum attainable resolution can be obtained by replacing the maximum number of pixels per inch in equation (3.68) in the

118   CHAPTER 3. WORKSTATION SUPPLY MODELS

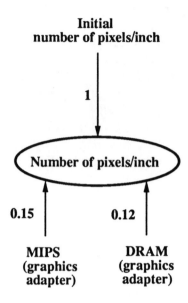

Figure 3.15: Relational diagram of the number of pixels per inch of a color CRT display.

resolution equation (3.71):

$$MaxRES_t = \frac{(HS_{in} * MaxPPI_t) * (VS_{in} * MaxPPI_t)}{10^6} \;\; (megapixels).$$
(3.76)

However, the maximum resolution is rarely the manufacturing standard because of difficulties encountered in deflecting the electron beams to penetrate the exact assigned hole in the metal shadow mask to hit the corresponding color phosphors. So the actual number of pixels per inch is slightly lower than the number of holes per inch in the metal shadow mask.

Resolution and Screen Hardware Drivers

Figure 3.15 illustrates, in a relational diagram, what effects the screen hardware drivers of the CRT have on the number of pixels per inch and the resolution, namely the speed of the graphics microprocessors (and/or microcontrollers) used and the capacity of the DRAMs installed. The

3.4. COLOR CRT DISPLAY SUPPLY MODEL

graphics microprocessors execute a series of instructions fed to them by the running application program. In turn, they access the display allocated DRAMs and send the fetched data through a chip that converts the digital signals to analog signals[33]. The levels of the analog signals set the intensities of the electron beams and the color shades of the pixels. The speed at which the analog signals are fed to the CRT sets the bandwidth. It is apparent that the attributes of the hardware drivers of the CRT have been pivotal in obtaining higher resolutions and a larger number of colors and, in turn, made the ICs technology-driving trends the CRT resolution's.

The top-of-the-line hardware drivers map each pixel to 24 bits of digital data, or 8 bits per color, and, since the digital bits are mapped onto analog signal levels, the actual number of colors that the screen can emit is equal to $2^{3*\#bits/color}$, in this case, 2^{24} or close to 17 million different colors [82].

The hardware drivers are referred to in the literature as graphics or video adapters. Most of the graphics microprocessors have been CISC architecture based [72]. Their relative performance can be measured in MIPS because the instructions are fairly simple and, in most cases, do not involve double precision floating point operations. The higher the MIPS rating, the faster the microprocessor executes the graphics instructions, and the faster the bits of each pixel get fed to the D/A converter to be displayed on the screen. Furthermore, as indicated in Figure 3.15, a higher DRAM capacity results in a cheaper cost per megabyte of DRAM, an increase in the number of DRAM chips that can be packaged on the graphics adapter, and a higher number of pixels per inch.

The effects of the accelerating, focusing, and deflecting apertures are not considered as factors of the number of pixels per inch behavior because their effects are cumulative, and there are no individual data trends relating them to the number of pixels per inch of color display. The cost per megapixel equation, to be presented later, will incorporate their effects as cost-decreasing factors in an exponentially decreasing component of the equation.

The number of pixels per inch model, illustrated in Figure 3.15, can

---

[33] The chip that converts the digital signals to analog is referred to as a D/A converter.

be expressed mathematically as follows:

$$PPI_t = PPI_0 * \left(\frac{MIPS_t}{MIPS_0}\right)^{0.15} * \left(\frac{MEMORY_t}{MEMORY_0}\right)^{0.12} \quad (pixels/inch) \tag{3.77}$$

where PPI_0 corresponds to the initial value of the number of pixels per inch. The values of the exponents, $(a_n s)$, in equation (3.77) were obtained by tuning the model to yield results to match the actual number of pixels per inch data provided in Table 3.8. All of the $|a_n|$s are ≤ 1, and each reflects an estimate of its relative importance to the behavior of the number of pixels per inch over time. For example, a 10 percent increase in the microprocessor's MIPS rating, which might incorporate an increase in its speed and a more efficient instruction set [see equation (3.20)], is coupled with an approximate increase of 1.5 percent in the number of pixels per inch [see equation (3.9)], because faster adapters can drive a larger number of screen pixels. Also, a 10 percent increase in the capacity of the adapter's DRAM is coupled with an approximate increase of 1.2 percent in the number of pixels per inch because more DRAMs enable more pixels to be stored during the refresh cycle.

3.4.6 Metal Shadow Mask Manufacturing Yield

As the hole pitch decreases, manufacturing the metal shadow mask becomes more intricate and more costly. Associated with the manufacturing of the shadow masks are their yields, which are directly proportional to the hole pitch values. Data on shadow mask production yields and mask defects per unit length are classified company information [16], so their actual values can only be estimated in their effects on the behavior of the cost per megapixel over time. In equations (3.78) and (3.79), these effects are captured in the magnitude and sign of the exponents. What follows is the mask's yield equation, which relates the number of mask production defects per unit length (MDPUL) to the diagonal size (DS) of the CRT:

$$MY_t = (MDPUL_t * DS_{in})^{\frac{-MDPUL_t}{4}} \tag{3.78}$$

where $MDPUL_t$ is expressed as:

$$MDPUL_t = MDPUL_0 * \left(\frac{HP_t}{HP_0}\right)^{-0.2} \quad (defects/inch) \tag{3.79}$$

where, as the hole pitch (HP) decreases, the number of mask defects increases and, consequently, the mask yield decreases.

3.4.7 Color CRT Display Cost

Now let's turn our attention to modeling the cost of the color CRT display. Since 1989, the prices of high-resolution color CRTs decreased by approximately 10 percent per year [2], spurred by an increase in the number of manufacturers, enhanced production techniques, and higher CRT production yields. Any time a higher resolution is sought, the entire system's specifications change. The ability to transfer existing technology to new designs gives the CRT-based displays an edge over other contemporary technologies. The total cost of a 19-inch color display is equal to the product of the resolution and the cost per megapixel:

$$TCM_t = RES_t * CMP_t \quad (\$). \tag{3.80}$$

The factors affecting the cost per megapixel of a color CRT display and the mathematical expression of the cost per megapixel are discussed below.

Cost per Megapixel

The overall cost per megapixel is a function of the quality of the display, its attributes, the precision of the design, the types of color phosphors used, the maximum bandwidth[34] it can handle, the hole pitch and the metal shadow mask yield, the graphics adapter, and the accelerating, focusing, and deflecting apertures. It is difficult to find a consistent display pricing strategy on the market because of the previously listed factors.

Figure 3.16 illustrates, in a relational diagram, the effects on the overall CRT cost per megapixel of the metal shadow mask production yields, the MIPS rating of the graphics adapter, and the amount of DRAM installed:

1. Metal shadow mask yield was chosen as a CRT technology-driving trend and, indirectly, a CRT cost-driving trend because a higher yield is translated in lower CRT manufacturing costs and, consequently, lower cost per displayed megapixel.

2. MIPS was chosen as a cost-per-megapixel-driving trend because, as the MIPS rating of the graphics microprocessor improves—

---
[34] Several monitors on the market offer varying operational bandwidths, due to the variety of graphics adapters available to drive the CRT.

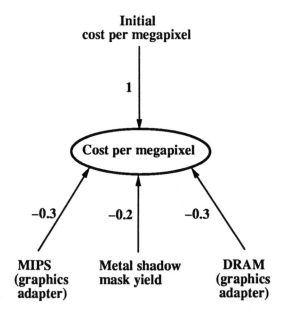

Figure 3.16: Relational diagram of the cost per megapixel of a color CRT display.

3.4. COLOR CRT DISPLAY SUPPLY MODEL

incorporating an increase in its speed and a more efficient instruction set [see equation (3.20)]—the price per MIPS decreases (see actual data in section 3.2 and ICs supply model results in section 4.3). Thus, if the refresh rate, the number of colors displayed, and the resolution of the CRT are assumed to remain constant, the cost of the graphics adapter and, consequently, the cost per megapixel decrease.

3. DRAM capacity was chosen as a cost-per-megapixel-driving trend because, as the DRAM capacity of the graphics adapter increases, the DRAM price per megabyte decreases (see actual data in section 3.2 and ICs supply model results in section 4.3). Thus, if the refresh rate, the number of colors displayed, and the resolution of the CRT are assumed to remain constant, the cost of the graphics adapter and, consequently, the cost per megapixel decrease.

The cost per megapixel model can be expressed mathematically as follows:

$$CMP_t = CMP_0 * \left(\frac{MY_t}{MY_0}\right)^{-0.2} * \left(\frac{MIPS_t}{MIPS_0}\right)^{-0.3} * \left(\frac{MEMORY_t}{MEMORY_0}\right)^{-0.3} * \theta_t \ (\$/MP) \quad (3.81)$$

where CMP_0 is equal to the value of equation (3.70) evaluated at t_0, MY represents the metal shadow mask production yield, $MIPS$ represents the number of million instruction per second that the graphics microprocessor can perform, $MEMORY$ represents the DRAM capacity installed on the adapter board, and t represents the year in which the cost per megapixel is calculated. Since most of the design yields and the effects of the apertures enhancements on the cost are kept as classified information [16], an exponentially decreasing factor, $\theta_t = 10^{-0.004*(t-t_0)}$, is used in equation (3.81) to capture these cost-diminishing effects over time. The values of the exponents, $(a_n s)$, in equation (3.81) were obtained by tuning the model to yield results to match the actual cost per megapixel data provided in subsection 3.4.2. All of the $|a_n|$s are ≤ 1, and each reflects an estimate of its relative importance to the behavior of the cost per megapixel over time. For example, a 10 percent increase in the metal shadow mask production yield leads to an approximate decrease of 2 percent in the cost per megapixel [see equation (3.9)], and similarly for the other factors of equation (3.81).

3.5 UNIX Operating System Supply Model

Software has always played a key role in the user's perception of a computer system. With more and more UNIX-operated workstations sold and with more UNIX supported applications developed, UNIX has become the operating system of choice for many large-scale network-distributed and multiprocessed applications [30]. The latest version of the UNIX operating system has more than 1 million lines of code [78]. With widespread support from the Open Software Foundation (OSF) and the engineering and business communities [30, 96], UNIX[35] has become one of the major operating systems in the computer industry.

This section is organized as follows: subsection 3.5.1 presents the UNIX development-from-scratch and porting time periods and costs for both 1980 and 1991; subsection 3.5.2 presents the assumptions and the terminology used in the development of the UNIX supply model; and subsections 3.5.3 through 3.5.7 present the mathematical formulation of the UNIX operating system supply model, including equations that capture the time periods and costs of both the development-from-scratch and the porting of UNIX over the 1980–1991 study period.

3.5.1 UNIX Development-from-Scratch and Porting Trends

The times required for UNIX development-from-scratch and porting are company-dependent [73]. Developing software involves creativity and experience, two intangible and immeasurable attributes. To facilitate model development, some experienced UNIX developers at Uniforum [73] were consulted about development-from-scratch and porting issues related to the UNIX operating system. Based upon these discussions, a model was formulated. The next two paragraphs report the UNIX development-from-scratch and porting time periods and costs data for 1980 and 1991. Unfortunately, data for the years between 1980 and 1991 were not available.

---

[35] The main attributes of UNIX that have enabled the computer workstation to achieve its current marketability are its optimal hardware resources utilization, single user multitasking, networking, and distributed computing support [78, 86].

3.5. UNIX OPERATING SYSTEM SUPPLY MODEL

1980 Development-from-Scratch and Porting: Time and Cost Data

Since UNIX was not as popular in the early 1980s as it is today, the knowledge and experience of UNIX software engineers were not as sophisticated as is needed today. From the UNIX development specialists we learned that the UNIX development-from-scratch time period was nearly 15 man-years in 1980 and that the porting time was approximately 1.5 man-years [73]. The demand for UNIX software engineers far exceeded their availability in the late 1970s, and they charged a higher coding[36] cost per hour. The same UNIX development specialists estimated that the average coding cost per hour was close to $70 in 1980 [73]. Based on the 1980 development-from-scratch and porting time periods and the average coding cost per hour, the development-from-scratch and porting costs[37] can be computed as follows:

$$DFSC_{1980} = 15 * (70 * 8 * 250) = 2,100,000 \ (\$) \qquad (3.82)$$
$$PC_{1980} = 1.5 * (70 * 8 * 250) = 210,000 \ (\$) \qquad (3.83)$$

where $DFSC_{1980}$ is the 1980 development-from-scratch cost of UNIX, PC_{1980} is the 1980 porting cost of UNIX, 8 is the assumed number of working hours per day, and 250 is the assumed number of working days per man-year.

1991 Development-from-Scratch Time and Cost Data

Several workstation manufacturers like DEC, HP, IBM, and Sun Microsystems, Inc., have developed their own proprietary UNIX versions and packaged them in their computer systems. If the capabilities of UNIX were developed from scratch in 1991, it has been estimated that it would take only about 10 man-years [73]. When one considers also that in 1991 a software developer commanded an average salary of $60,000 per year, and that the yearly job benefits and overhead totaled approximately $40,000, the cost of developing the UNIX operating system from scratch would drop to nearly $1 million [73]. Of this cost, the storage tape on which the UNIX software is kept costs no more than $10 [73].

---

[36] Coding is the job performed by the software engineer while developing the operating system's code; most of the UNIX code is written in the C programming language.

[37] The development-from-scratch and porting costs reported are estimates, and the actual costs depend on the development company's overhead and its policy regarding the number of software prototypes, or beta-versions, that it tests before final release.

1991 Porting Time and Cost Data

UNIX porting costs are not as extensive as the development-from-scratch costs; nevertheless, it takes about one man-year to port UNIX to a different hardware platform [73] and costs close to $100,000.

It is apparent from the foregoing that the UNIX development-from-scratch and porting time period and cost trends are decreasing. These trends reflect the much increased popularity of UNIX since the market introduction of the computer workstation and indicate how widespread porting and building new applications on top of the UNIX operating system is today.

3.5.2 Model Assumptions and Terminology

The operating system supply model incorporates equations capturing the time periods and costs required for the development-from-scratch and porting of the UNIX operating system. It costs no more than $10 to package a copy of the UNIX software in a magnetic tape, so copying and packaging costs are negligible and will not be considered as an integral part of the supply model.

The main costs, after incurring the development-from-scratch or porting costs, involve software enhancements. Each company adds several enhancements to its own version to suit its machines and, depending on the demand for that company's machines, the price of the OS could vary from $800 to $2,000 [87]. Before presenting the mathematical formulation of the model, several assumptions are listed. This is followed by a list defining the variables and the parameters used in the model.

Model Assumptions

In modeling the supply of developed-from-scratch or ported UNIX software, the following assumptions are used:

1. The number of UNIX-based machines sold is increasing during the period of study [85, 86, 87].

2. The number of UNIX development-from-scratch and porting consultants is increasing during the period of study [73].

3. The average UNIX coding cost per hour is decreasing during the period of study [73].

Definitions and Terminology

In presenting the model's equations, several parameters and variables are introduced, and they are defined as follows:

0	time index for the first year of the study period
t	time index
SA	software attributes index
WHA	workstation hardware attributes index
$MIPS$	number of million instructions per second
$MEMORY$	capacity of a DRAM, in megabytes
DB	display bandwidth, in megaHertz
DR	hard disk data rate, in megabytes per second
UDS	number of UNIX development specialists
$DUDS$	demand for UNIX development specialists
$ACCH$	average coding cost per hour, in dollars per hour
$DFST$	UNIX development-from-scratch time period, in years
PT	UNIX porting time period, in years
$DFSC$	UNIX development-from-scratch cost, in dollars
PC	UNIX porting cost, in dollars

There are no standard software development tools and techniques for use by all software developers as of 1991. More than 50 books have been written on software development techniques [71] and the problems remain the same: how to make an *efficient* transition from the structural specifications of a software project to its actual completion, and how to assign a cost to a project from its structural specifications. The current software development paradigms do not allow these questions to be fully answered. What the operating system supply model attempts to capture is the overall effect of improved coding machines' hardware attributes on the time required to satisfy the structural specifications and on the cost associated with meeting these specifications.

The UNIX supply model is presented as follows: in subsection 3.5.3 the UNIX development-from-scratch time period is presented; in subsection 3.5.4 the UNIX porting time period is presented; in subsection 3.5.5 the software attributes index is described and an assumed index behavioral trend introduced; in subsection 3.5.6 the workstation[38] hardware attributes index is described with all its factors and an index behavioral

---

[38] Assuming that a computer workstation is used to develop the software code.

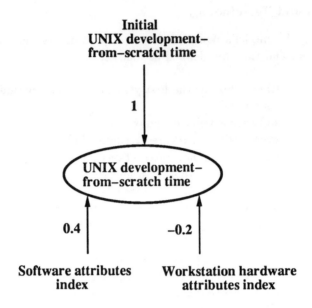

Figure 3.17: Relational diagram of the UNIX development-from-scratch time period.

equation introduced; and in subsection 3.5.7 the UNIX development-from-scratch and porting costs are computed.

3.5.3 UNIX Development-from-Scratch Time Period

Figure 3.17 illustrates, in a relational diagram, how software and hardware attributes affect the development-from-scratch time period of the UNIX operating system:

1. Software attributes index was chosen as a UNIX development-from-scratch time-driving trend because, as the software attributes increase, the software specifications increase and, usually, the time required to meet all of them increases.

2. Workstation hardware attributes index was chosen as a UNIX development-from-scratch time-driving trend because, as the hard-

ware attributes of the coding machine[39] become more advanced, the workstation's response time usually improves and, consequently, the software development task becomes more efficient.

The diagram in Figure 3.17 can be expressed mathematically as:

$$DFST_t = DFST_0 * \left(\frac{SA_t}{SA_0}\right)^{0.4} * \left(\frac{WHA_t}{WHA_0}\right)^{-0.2} \quad (year) \quad (3.84)$$

where $DFST_0$ represents the initial UNIX development-from-scratch time period. Equation (3.84) reflects how the UNIX development-from-scratch time period increases as the UNIX software attributes index (SA) increases, and decreases as the development workstation's hardware attributes index (WHA) increases. The values of the exponents, (a_ns), in equation (3.84) were obtained by tuning the model to yield results to match the UNIX consultant estimated development-from-scratch trend provided in subsection 3.5.1. All of the $|a_n|$s are ≤ 1, and each reflects an estimate of its relative importance to the behavior of the development-from-scratch period over time. For example, a 10 percent increase in the software attributes index leads to an approximate increase of 4 percent in the development-from-scratch time period [see equation (3.9)], and a 10 percent increase in the workstation hardware attributes index leads to an approximate decrease of 2 percent in the development-from-scratch time period.

3.5.4 UNIX Porting Time Period

The UNIX porting time period is affected by the same factors associated with the development-from-scratch of UNIX, and Figure 3.18 is the relational diagram showing how these factors might affect porting time. The diagram in Figure 3.18 can be expressed mathematically as:

$$PT_t = PT_0 * \left(\frac{SA_t}{SA_0}\right)^{0.4} * \left(\frac{WHA_t}{WHA_0}\right)^{-0.2} \quad (year) \quad (3.85)$$

where PT_0 represents the initial UNIX porting time period. The explanation of how each of equation (3.85)'s factors and their exponents affect the porting time is similar to that provided for equation (3.84) in the previous subsection.

---

[39] Computer workstations are usually used in the software development process.

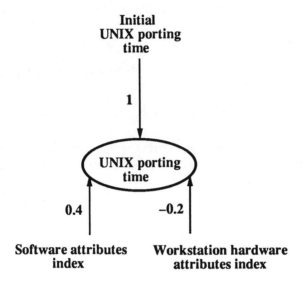

Figure 3.18: Relational diagram of the UNIX porting time period.

3.5.5 Software Attributes Index

The software attributes of the UNIX operating system depend on the user's perception and, as a result, they are often difficult to measure. These attributes include the user friendliness of the software, coupled with its compatibility, networkability, portability, and efficiency. For modeling purposes, it is assumed that the index of all the software attributes presented earlier is increasing at a rate of 5 percent a year, or mathematically:

$$SA_t = SA_0 * 1.05^{t-1980}. \tag{3.86}$$

It does not matter what the initial value SA_0 is because only the relative change in the values of the software attributes index is relevant in equations (3.84) and (3.85).

3.5.6 Workstation Hardware Attributes Index

Most computer hardware attributes are tangible entities, and their performance can be measured over time. Of these measurable attributes,

3.5. UNIX OPERATING SYSTEM SUPPLY MODEL

the following were selected as potentially relevant to expressing a coding machine's capabilities:

1. The MIPS rating of the workstation's integer unit was chosen because it reflects the processor's speed and architecture, and, partly, the workstation's throughput.

2. The DRAMs capacity was chosen as a hardware attribute because denser DRAMs enable most of the software developer's code to remain in main memory during execution, thereby increasing the machine's response time and, in turn, reducing the development time.

3. The CRT display bandwidth was chosen as a hardware attribute because it affects the size of the screen and, ultimately, its resolution. (It has been shown that larger screens improve the software development environment and make the coding process more efficient, and so reduce the development time.)

4. The hard disk data rate was chosen as a hardware attribute because it reflects the hard disk data density and the rate at which the data are fed back to the suspended program, while waiting on the data from the disk. (A fast data rate increases the disk's response time and, consequently, improves the efficiency of the code development process.)

All the above listed workstation hardware attributes are calculated in the ICs, magnetic hard disk, and CRT display supply models already reported in previous sections of this chapter. Use of these attributes integrates the models into one simple formulation. Figure 3.19 illustrates in a relational diagram the effects of the previously presented attributes on the overall index of the workstation's hardware attributes. The diagram in Figure 3.19 can be expressed mathematically as follows:

$$WHA_t = WHA_0 * \left(\frac{MIPS_t}{MIPS_0}\right)^{0.25} * \left(\frac{MEMORY_t}{MEMORY_0}\right)^{0.35} * \left(\frac{DB_t}{DB_0}\right)^{0.15} * \left(\frac{DR_t}{DR_0}\right)^{0.25}. \quad (3.87)$$

The initial value WHA_0 of the workstation hardware attributes index is not relevant because only the relative change of the index's values matters in equations (3.84) and (3.85). All of the a_ns add up to 1, and each reflects an estimate of its relative importance to the behavior of the

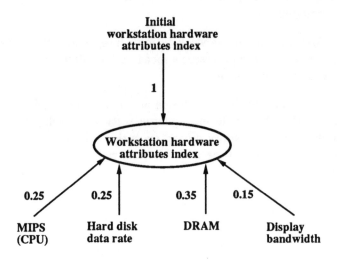

Figure 3.19: Relational diagram of the workstation hardware attributes.

hardware attributes index over time. For example, a 10 percent increase in the hard disk data rate leads to an approximate increase of 2.5 percent in the hardware attribute index [see equation (3.9)], and similarly for the rest of the factors of equation (3.87).

3.5.7 UNIX Development-from-Scratch and Porting Costs

As shown earlier, the development-from-scratch and porting processes have associated with them long periods of coding time. The coding time spent has, in turn, an associated cost. This coding cost has decreased over the years because of the increased demand and supply of UNIX development specialists [73]. It is expressed mathematically as:

$$ACCH_t = ACCH_0 * \left(\frac{UDS_t}{UDS_0}\right)^{-1} * \left(\frac{DUDS_t}{DUDS_0}\right)^{1} \quad (\$/hour) \qquad (3.88)$$

where $ACCH_0$ is chosen to reflect the initial average coding cost per hour at t_0. It is assumed that the number of UNIX development specialists and their market demand have been increasing yearly by 10 percent and 7 percent, respectively, since 1980, due to the proliferation of UNIX-based workstations in the private and the public sectors. The UNIX development specialists trend is expressed mathematically as follows:

$$UDS_t = UDS_0 * 1.1^{t-1980} \qquad (3.89)$$

and the market demand for UNIX development specialists is expressed as:

$$DUDS_t = DUDS_0 * 1.07^{t-1980}. \qquad (3.90)$$

The values of UDS_0 and $DUDS_0$ in equations (3.89) and (3.90) are immaterial because equation (3.88) takes into consideration only the relative increase in the number of UNIX specialists and their demand and (not their actual numerical values; t represents the year in which the number of UNIX development specialists and the demand for such specialists are calculated. From equations (3.84), (3.85), and (3.88), the development-from-scratch and porting costs of the UNIX operating system can be computed as follows:

$$DFSC_t = ACCH_t * (8 * 250 * DFST_t) \text{ (\$)} \qquad (3.91)$$
$$PC_t = ACCH_t * (8 * 250 * PT_t) \text{ (\$)} \qquad (3.92)$$

where the 8 * 250, or 2,000, is equal to the number of working hours in one calendar man-year [73].

3.6 Workstation Assembly Model

Assembling a workstation has become a global effort. Most workstation assembly plants ship their raw materials and components from all over the world, and the assembled workstations are, in turn, shipped to many destinations worldwide. To be competitive, a workstation manufacturer must develop a mechanism to minimize its costs on a global scale, taking into consideration all the transportation, labor, and materials costs, the capacities of each plant's productive units, and the market demand each plant's output can satisfy. A useful conceptualization by which the manufacturing activity can be optimized is a process model, where the assembly plant is considered as a network with several processing stages, each stage with its specific input requirements and output levels. Each productive unit corresponds in the model to an assembly node with throughput bounds. Increasing the throughput bound requires allocation of investment capital by the manufacturer.

The workstation assembly network is illustrated in Figure 3.20, where components—CPUs, DRAMs, magnetic hard disks, CRT displays, and UNIX operating systems, and so on—are assembled into a final product. But each of the components is itself an intermediate product. These first intermediate products are the CPU boards, the DRAM boards, the magnetic hard disks with their interfaces to the computer systems, the CRT displays with their interfaces to the computer systems, and the UNIX operating systems with their installation kits ready for loading. At a second intermediate stage, a CPU board, a DRAM board, a magnetic hard disk with its interface, and an electric power supply are assembled into a workstation box. At a third intermediate stage, the operating system is loaded in the hard disk, and the keyboard, mouse, and display with its interface are connected to the box's corresponding external ports[40]. The assembled workstation goes through a final burn-in stage before it is ready for shipping to the distribution centers. Each of the assembly points in the network has a capacity limit that can be expanded only by increasing the plant's space, machinery, raw materials inventory, and labor.

The developed linear process model provides an optimal allocation of inputs among the different activities to minimize the assembly costs over several intervals[41] of time. The model is a simple illustration of the workstation assembly process, and only one manufacturing plant is considered, with its market demand located in its vicinity (that is, no workstations transportation costs are considered). Minimizing the costs is the model's objective function. The cost minimization is constrained by internal capacity limitations and the market requirement for finished workstations that must be satisfied. Although no import, export, plant expansion, and government quota restrictions are included, conceptually there is no difficulty in adding them. There is extensive literature on the use of process models to analyze problems of the type considered here. For more details, see Kendrick and Stoutjesdijk [47] or Kendrick, Meeraus, and Alatorre [46].

The presentation of the workstation assembly model will proceed as follows: first, in subsection 3.6.1 the workstation assembly steps are described; second, in subsection 3.6.2 the assumptions and terminology

---

[40] A port is a physical connection to the machine's internal hardware and serves as a communication gate between the processing boards and the peripherals.

[41] An interval is usually one year because the available data are provided annually.

3.6. WORKSTATION ASSEMBLY MODEL

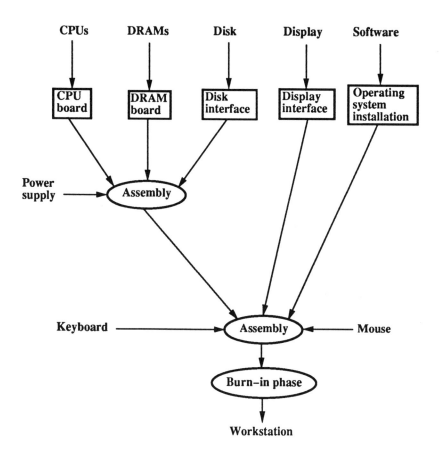

Figure 3.20: Illustration of a workstation assembly network.

used in the development of the workstation assembly model are presented; and third, in subsection 3.6.3 the model is formulated.

3.6.1 Assembly Steps

The assembly process in Figure 3.20 can be described in five steps:

1. The components meeting the specifications of the configured machine are prepared at the machine's assembly site.

2. The CPU and the DRAM chips are mounted on the printed circuit boards—that is, the processing board and the main memory (DRAM) board.

3. The processing board, the main memory board, the hard disk storage with its interface, and the electric power supply are grouped and tested for correct assembly.

4. If the already assembled hardware is functional, the operating system is loaded into the hard disk, and the keyboard, mouse, and display with its interface are connected to the machine.

5. A burn-in phase takes place where the functionality of the workstation is tested for a period of 10 to 24 hours.

3.6.2 Model Assumptions and Terminology

In formulating the linear process model the following assumptions were used:

1. The arcs in the network are directed.

2. The input amount of each component in the assembly process is constant during one time period.

3. The price of each component is constant during one time period.

4. Some of the hardware components are produced outside the United States. However, due to the lack of data on the geographical locations of the components manufacturers, it will be assumed that all the components needed for assembly will be available at the workstation's assembly site. The cost of transporting these components to the assembly plant will be incorporated in their prices.

3.6. WORKSTATION ASSEMBLY MODEL

5. Each CPU die incorporates an integer unit (IU), a floating point unit (FPU), a memory management unit (MMU), and a cache memory. The price of the CPU reflects the price of the set.

6. Only one assembly plant is considered.

7. The demand is assumed to be concentrated in the vicinity of the production plant; no workstation's transportation costs are considered.

8. No plant expansion costs are considered, and the market demand for the workstations produced from the plant remains constant at a value less than or equal ($\leq$) to the plant's production capacity.

9. The monetary discount rate is assumed to be 5 percent, compounded yearly. The discount rate does account for the inflation of prices during the study period.

Definitions and Terminology

In presenting the linear process model's equations, several parameters and variables are introduced, and they are defined as follows:

t	index to the time periods T
c	index to the commodities C, CR, CI, and CF
p	index to the set of processes P
m	index to set of productive units M
DR	discount rate, in percent
A_{cp}	input-output coefficients matrix
B_{mp}	plant capacity utilization matrix
K_{mt}	capacities of the productive units matrix
$DDIST$	demand distribution matrix, in percent
$PRICES_{ct}$	price matrix of the components and raw materials
z_{pt}	process level variable
x_{ct}	amount of final products shipped variable
u_{ct}	amount of materials purchased variable
$RMATC_t$	cost variable of the components and raw materials

3.6.3 Model Formulation

The *objective* function in the linear model is to minimize the present value of the total workstation components and raw materials costs, where future costs are discounted using the market rate:

$$\text{MIN} \sum_{t=1}^{T} (1 + \frac{DR}{100})^{(1-t)} * RMATC_t. \qquad (3.93)$$

This model deals with the assembly aspect of workstations only. For instance, no transportation costs are included in equation (3.93) because we assumed that the demand is in the vicinity of the production plant. No expansion costs are included because we assumed that the demand is constant over the period of study.

The workstations components and raw materials costs equations are equal to the products of the prices provided in Table 4.16 and the amount of components and raw materials required in the production process:

$$RMATC_t = \sum_{c=1}^{CR} PRICES_{ct} * u_{ct} \quad \text{for } t = 1 \cdots T. \qquad (3.94)$$

Since no component imports or product exports are considered, the raw materials balance constraints at the production plant are:

$$\sum_{p=1}^{P} A_{cp} * z_{pt} \geq -u_{ct} \quad \text{for } c = 1 \cdots CR, \; t = 1 \cdots T. \qquad (3.95)$$

The intermediate materials balance constraints are:

$$\sum_{p=1}^{P} A_{cp} * z_{pt} \geq 0 \quad \text{for } c = 1 \cdots CI, \; t = 1 \cdots T. \qquad (3.96)$$

And the final materials balance constraints are:

$$\sum_{p=1}^{P} A_{cp} * z_{pt} \geq x_{ct} \quad \text{for } c = 1 \cdots CF, \; t = 1 \cdots T. \qquad (3.97)$$

The capacity constraints at the assembly plant require that the plant's output cannot exceed the capacity of its production units:

$$\sum_{p=1}^{P} B_{mp} * z_{pt} \leq K_{mt} \quad \text{for } m = 1 \cdots M, \; t = 1 \cdots T. \qquad (3.98)$$

3.6. WORKSTATION ASSEMBLY MODEL

The demand constraints guarantee that the number of workstations produced from the plant satisfy the market demand:

$$x_{ct} \geq 10,000 * \frac{DDIST}{100} \quad \text{for} \quad c = 1 \cdots CF, \, t = 1 \cdots T \quad (3.99)$$

and since we are assuming the demand is generated by a single market, *DDIST* is equal to 100 percent.

There are several non-negativity constraints, including:

$$z_{pt}, \, x_{ct}, \, u_{ct} \geq 0 \quad \text{for} \quad c = 1 \cdots C, \, p = 1 \cdots P, \, t = 1 \cdots T. \quad (3.100)$$

The linear process model was coded in GAMS [11], and the BDMLP[42] solver on an IBM 3081KX VM/XA mainframe was used to solve it.

---

[42] BDMLP is a linear programming solver.

4
Model Behavior and Sensitivity Results

The workstation component supply models presented in Chapter 3 are discrete event simulation models. Each model was designed and tuned to capture certain component trends over time. This chapter illustrates the dynamic behavior of the component supply models. This is not a validation of the models. A true validation, in a scientific sense, is only possible by using the models to test hypotheses. In other words, a model prediction needs to be made (a hypothesis formed), and if the prediction turns out to be true, the model passes the validation test.

More important than model validity in the strict scientific sense is how useful the model is at helping us to understand and gain insight into past and possible future trends. To illustrate the models' utility in this regard, we present a base case scenario of future trends for each of the workstation components and their assembly into completed workstations. This is followed by sensitivity analyses which make clear the strength of the models' linkages. As will be seen, this set of analyses indicates that the supply models developed in Chapter 3 can yield practical insights into the relationships among a workstation's components, capabilities, and costs.

A careful examination of how the model results over a past period of time compare with actual market data over the same historical period is presented. This discussion, however, simply shows how well the supply models of Chapter 3 were tuned to capture the historical behavior of the data. A stronger form of comparison is simply not possible with the current models and the actual available data. The closer the models' results are to the actual market data, the better tuned the models are, unless the actual market data represent company-dependent decisions that the models cannot capture.

The supply models were tuned to capture unit market price trends of the various workstation components and the behavior of certain component capabilities over time. The unit price results of the component supply models include the CPU and DRAM prices, the price per megabyte of magnetic hard disk, the price per megapixel of color CRT display, and the UNIX porting and development-from-scratch prices. The component capabilities include the operational speed of the microprocessor, the capacity of the main memory, the capacity of the magnetic hard disk, the size and resolution of the color CRT display, and the functionality of the operating system.

Most of the market data collected reflect the attributes of the components during the 1980s; consequently, the components supply models[1] were simulated and their results compared to actual data over the 1980–1991 period. The simulation code of the component supply models was written in C, and an IBM RISC System/6000 POWERserver/530 was used to obtain the simulation results.

This chapter is organized as follows:

- Section 4.1 presents the cost markup percentage used in the supply models to obtain the component prices, and illustrates the difference between the single-unit and the bulk price of a component.

- Section 4.2 lists the component supply models' input parameters and the rates of change of each of the components' physical characteristics, capabilities, and price trends, if available.

- Sections 4.3 through 4.6 present and compare the results of the supply models and the actual market data of the IC components, the magnetic hard disk, the color CRT display, and the UNIX operating system.

---

[1]The supply models reflect only the attributes of the components already on the market, not those available in the manufacturers' laboratories. Most of the future ICs and computer systems products are provided in the IEEE International Solid-State Circuits Conference–ISSCC Digest of Technical Papers and the IEEE COMPCON, respectively.

- Section 4.7 presents the workstation assembly model inputs and results.

- Section 4.8 performs sensitivity analyses of the workstation components and assembled workstations models to the ICs' feature size and the number of silicon wafer defects per unit area.

4.1 Component Cost, Single Unit Price, and Bulk Price

Since the supply models provide single-unit component costs, not prices, as their output, all of the cost results were marked up by 200 percent to compare them with the available market price data [33]:

$$Single\ Unit\ Price_{model} = Single\ Unit\ Cost_{model} * (1+200\%)\ (\$). \quad (4.1)$$

The 200 percent cost markup percentage includes the gross margin and the average discount percentages that the manufacturers add to their products' costs[2].

Most manufacturers sell their products in bulk, in quantities larger than or equal to 10,000 units. The bulk unit price is usually equal to half the single-unit list price [16]:

$$Bulk\ Unit\ Price = \frac{Single\ Unit\ Price}{2}\ (\$). \quad (4.2)$$

Since the actual data collected are single-unit list prices, though, the results of the models are presented in the following sections as single-unit list prices.

4.2 Component Supply Model Inputs

Each of the component supply models has a set of input parameters, the values of which can be changed interactively while executing the simulation program. As presented in Chapter 3, certain CPU and DRAM attributes of the ICs model, like the MIPS of CISC CPUs and the capacity of DRAMs, were incorporated as factors influencing the attributes of

---
[2]Some manufacturers mark up their components' costs by as much as 300 percent [33].

144   CHAPTER 4. MODEL BEHAVIOR AND SENSITIVITY RESULTS

other models; the inputs of the ICs supply model are, then, a part of the input parameters of all the supply models.

This section presents the input parameters of the component supply models in subsection 4.2.1, and lists the rates of change of the components physical characteristics trends in subsection 4.2.2 and the rates of change of the components capabilities and price trends in subsection 4.2.3.

4.2.1 Model Input Parameters

The list below presents the values of the input parameters of the component supply models for the first year of the study period, or for the first year of the component's market introduction. For instance, the RISC architecture was widely marketed in 1987 [85], and the 19-inch computer color CRT display was introduced in 1985 [16]; their input parameters are given for the years 1987 and 1985, respectively. Even though the study period spans from 1980 until 1991, the RISC CPUs and the color CRT display models will be simulated only from 1987 to 1991, and from 1985 to 1991, respectively.

Simulation Period
  Starting year = 1980
  Study Period = 1980–1991
ICs Supply Model Inputs
  1980 – Wafer cost = \$350
  1980 – Wafer yield = 90 percent
  1980 – Number of silicon wafer defects per cm^2 = 2.5 defects/cm^2
  1980 – Number of test-dies per wafer = 2 dies
  1980 – Testing Cost per Hour = \$210/hour
  1980 – Average die test time = 15 seconds
  1980 – CISC CPUs speed per cm^2 = 25 megaHertz/cm^2
  1987 – RISC CPUs Speed per cm^2 = 28 megaHertz/cm^2
  1980 – DRAM density = 0.025 megabyte/cm^2
Magnetic Hard Disk Supply Model Inputs
  1980 – Head-actuator setup width = 1.25 cm
Color CRT Display Supply Model Inputs
  1985 – Screen refresh rate = 70 Hertz
  1985 – Number of metal shadow mask defects per inch = 1 defect/inch
UNIX Operating System Supply Model Inputs

1980 – Average coding cost per Hour = \$70/hour
1980 – UNIX porting period = 1.5 man-years
1980 – UNIX development-from-scratch period = 15 man-years

4.2.2 Components' Physical Characteristics Trends

The list below presents the rates of change of the components' physical characteristics trends during the study period, if available.

ICs Supply Model
CISC CPUs die areas: 7.53 percent per year exponential growth
DRAMs die areas: 7.53 percent per year exponential growth
Feature size: 5.5 percent per year exponential decline
Number of masking levels: 8 percent per year exponential growth
Silicon wafer diameter: 2.52 percent per year exponential growth
Magnetic Hard Disk Supply Model
Magnetic bit cell length: 4.9 percent per year exponential decline
Magnetic disk track pitch: 6.6 percent per year exponential decline
Magnetic medium thickness: 2.95 percent per year exponential decline
Magnetic head gap spacing: 4 percent per year exponential decline
Head-medium spacing: 5.5 percent per year exponential decline
Head-actuator setup width: 4 percent per year exponential decline
Color CRT Display Supply Model
Metal shadow mask hole pitch: 4 percent per year exponential decline

4.2.3 Components' Capabilities and Price Trends

The list below presents the rates of change of the components' capabilities and price trends during the period of study, if available.

ICs Supply Model
CISC CPUs IPC: 11 percent per year exponential growth
RISC CPUs IPC: 6.3 percent per year exponential growth
Magnetic Hard Disk Supply Model
Price per megabyte: 8.57 percent per year exponential decline
Areal density: 13.8 percent per year exponential growth
Color CRT Display Supply Model
Price per megapixel: 8.5 percent per year exponential decline

Table 4.1: Model results and actual market data of die yields for certain die areas in 1989. Source: [33].

ICs : Die Yields		
1	2	3
Die Area (cm^2)	Die Yield (Model Results) (%)	Die Yield (Actual Data) (%)
0.0625	77.3	78
0.2601	49.8	46
0.5776	27.3	22
1.0404	13.4	10
1.6129	6.6	5
2.3104	3.3	3
3.1684	1.6	2
4.1209	0.9	1

4.3 ICs: Model Results and Actual Market Data

The following section presents the ICs supply model results and compares them to the actual ICs market data presented in section 3.2. In subsection 4.3.1 the model results and the estimated market data of the ICs die yields are presented and compared; in subsections 4.3.2, 4.3.3, and 4.3.4 CISC CPUs, RISC CPUs, and DRAMs model results and actual market data are presented and compared.

4.3.1 ICs Die Yields

Column 1 of Table 4.1 lists some common die areas of manufactured ICs, and columns 2 and 3 present the corresponding 1989 model results on die yields and actual market data[3] for the various die sizes. The values in columns 2 and 3 differ, on average, by less than 16 percent.

Most of the die yields data are classified as confidential information

---

[3]The market data were estimated in Chapter 2 of Hennessy and Patterson [33], and were presented in section 3.2 of this book.

by the manufacturers because they reflect their positions on the learning curves and their percentages of profit. If made public, the die yields information can be used by competitors to create strategies to gain market share from the manufacturer whose die yields information was disclosed.

4.3.2 ICs: CISC CPUs Model Results and Actual Market Data

The following paragraphs compare the model results and the actual market data of CISC microprocessors. MIPS ratings, prices, price per MIPS, and speed per cm^2 of CISC CPUs are provided.

CISC CPUs MIPS and Prices

Table 4.2 presents a comparison of selected model results with actual data of Intel CISC microprocessors. Only Intel's data were considered because Motorola's CISC CPUs price data were not available. For each Intel CPU die area, the MIPS rating and the price are computed, and compared to the data already presented in Tables 3.2 and 3.3 of Chapter 3. The MIPS[4] model results and the actual data in columns 4 and 5 of Table 4.2 differ, on average, by less than 33 percent.

Some of the CPU model price results and actual price data in Table 4.2 differ significantly for certain CPUs. The actual market price of the i80286 in 1982 differs from the model result by almost $300. This is probably because, at the time, the Intel processors had gained wide acceptance for use in the IBM PC. What enhanced the i80286's acceptance in 1982 was IBM's decision to make public its bus and Intel-based processor board architectures to compete with and perhaps capture
Apple Computer's market share of PCs. As a result, the demand for the i80286s increased and Intel took a monopolistic stance, charging a sizable markup—larger than 200 percent—since it was the sole producer of that chip. By 1985, when the first i80386s were introduced, the model results and the actual data compare much more closely.

The difference between the 1991 actual market price data of the i8086/286/386DX ICs and the model price results can be attributed to

---

[4]The MIPS data and results reflect the architectural and production improvements to the CPU itself only, and not the entire chip set where it is used. Hence, the MIPS rating indicates solely the relative performance difference from other ICs in its class.

Table 4.2: Model results and actual market data on MIPS ratings and prices of Intel CISC CPUs. Sources: [42, 43, 49].

ICs: CISC CPUs						
1	2	3	4	5	6	7
Year	CPU Model #	Area (cm^2)	MIPS (Model Results)	MIPS (Actual Data)	Price (Model Results) ($)	Price (Actual Data) ($)
1982	i80286	0.60	0.82	2	67	360
1985	i80386DX	1.13	3.73	3	330	299
1989	i80486DX	1.65	17.95	11.40	552	700
1991	i8086	0.28	0.38	0.33	35	1.50
1991	i80286	0.60	1.98	2	61	7
1991	i80386DX	1.13	9.11	8.30	321	166
1991	i80486DX	1.65	32.60	35	558	588

the fact that the old technologies were nearing obsolescence. As new technologies are introduced, the demand for old ones diminishes, forcing their prices down. After a while, the i8086/286/386DX market prices were even depressed below a fully marked up production cost. Moreover, in 1991 Intel wanted to sell its stock of old CPUs to reduce their high inventory costs, which stimulated more price cuts.

CISC CPUs Price per MIPS

Table 4.3 compares the price per MIPS model results with the actual price per MIPS data of CISC CPUs. The entries in Table 4.3 were obtained by dividing the price of each CPU in Table 4.2 by its MIPS rating. The model results and actual market data reveal a trend of decrease in the price per MIPS due to architectural design enhancements of CISC CPUs and higher manufacturing yields. The average error is less than 55 percent.

4.3. ICS: MODEL RESULTS AND ACTUAL MARKET DATA

Table 4.3: Model results and actual market data on the price per MIPS of Intel CISC CPUs. Sources: [42, 43].

ICs: CISC CPUs		
1	2	3
Year	Price per MIPS (Model Results) ($)	Price per MIPS (Actual Data) ($)
1982	81.9	180
1985	88.5	99.6
1989	30.6	61.4
1991	17.1	16.8

CISC CPUs Speed per cm^2

Table 4.4 lists the model results and the actual market data on the operational speeds per unit die area of CISC CPUs. (Unavailable data are denoted by n.a.) The average difference between the actual data and the model results is less than 14 percent. The model results range from 25 to 40.4 megaHertz per cm^2 in the 1980–1991 period, while the market data are scattered in the 24–40.8 megaHertz per cm^2 range during the same period. Some of the values in columns 2 and 3 of Table 4.4 differ significantly due to the different chip designs and layouts used by the manufacturers during the period of study. As mentioned in Chapter 2, a simpler CPU design always improves the speed because the chip layout becomes easier and the inner-chip communication time decreases due to the enhanced and optimized layout.

4.3.3 ICs: RISC CPUs Model Results and Actual Market Data

The following paragraphs compare the model results and the actual market data of RISC microprocessors. Where available, MIPS ratings, prices, and speed per cm^2 of RISC CPUs are provided.

Table 4.4: Model results and actual market data on the speed per cm^2 rating of CISC CPUs. Sources: Intel data [42, 43, 49], Motorola data [62, 72, 85].

ICs: CISC CPUs		
1	2	3
Year	Speed per cm^2 (Model Results)	Speed per cm^2 (Actual Data)
1980	25.00	n.a.
1981	25.97	n.a.
1982	27.29	30.3
1983	28.35	24.4
1984	29.76	n.a.
1985	30.91	33.8
1986	32.42	n.a.
1987	33.97	40.8
1988	35.28	29.2
1989	36.94	n.a.
1990	38.66	30.3
1991	40.43	40

RISC CPUs MIPS and Prices

As mentioned earlier, the ICs model simulation period for the RISC[5] CPUs starts in 1987, instead of 1980, and ends in 1991. Hewlett-Packard's Precision Architecture (HP-PA) RISC CPUs data are compared with the model results because their die areas were the only actual data available. The MIPS ratings provided in Table 4.5 for the HP-PA CPUs are those of the integer units of the processing chip sets used in the HP workstations or servers [22, 38]. The actual MIPS ratings data (column 5) are increasing at a faster rate than the model results (column 4). This may be due to our lack of knowledge of the actual HP's manufacturing processes. One explanation for such an improvement in performance is related to the transfer of HP's mainframe technologies to the workstation[6]. No HP RISC CPUs price data were available because most of the HP chips are proprietary.

RISC CPUs Speed per cm^2

Table 4.6 lists the model results and the HP-PA market data on the operational speeds per unit die area of RISC CPUs. Again, the model results are increasing at a slower pace than those of the market data for the same reasons as given in the previous paragraph. The model results show an increase from 28 to 33.96 megaHertz per cm^2 in the four-year period, while the HP-PA CPU operational speed per unit die area actually more than doubled in a period of two years from 15.3 to 33.67 megaHertz per cm^2.

4.3.4 ICs: DRAMs Model Results and Actual Market Data

The following paragraphs compare the model results and the actual Motorola market data of DRAMs. Capacities, capacity per cm^2, prices,

---

[5]The RISC CPUs die areas and operational speeds are larger than the CISC's because the RISC architecture is simpler than CISC's, and the RISC CPU layout is much easier to manufacture and has a higher yield than a similar die area CISC CPU yield.

[6]Since the workstation market has been eating away at the mainframe's market share for the last ten years, HP might have deemed it necessary to capture a large share of the fastest growing sector of desktop and networkable computers—the workstation sector—while foregoing larger short-term profits that could have been made from selling mainframes.

Table 4.5: Model results and actual market data on MIPS ratings and prices of HP RISC CPUs. Sources: [7, 22, 38, 52, 54, 91, 105].

ICs: RISC CPUs						
1	2	3	4	5	6	7
Year	System	Die Area (cm^2)	MIPS (Model Results)	MIPS (Actual Data)	Price (Model Results) ($)	Price (Actual Data) ($)
1987	HP9000/825	1.33	23	9	404.91	n.a.
1988	HP9000/835	1.58	32	14	507.36	n.a.
1989	HP9000/845	1.96	49	22	787.80	n.a.
1991	HP9000/730	1.96	71	76	804.81	n.a.

Table 4.6: Model results and actual market data on the speed per cm^2 rating of RISC CPUs. Sources: HP data [7, 22, 38, 52, 54, 91, 105], Sun data [85, 87].

ICs: RISC CPUs		
1	2	3
Year	Speed per cm^2 (Model Results)	Speed per cm^2 (Actual Data)
1987	28.00	n.a.
1988	29.64	n.a.
1989	31.03	15.3
1990	32.47	n.a.
1991	33.96	33.67

Table 4.7: Model results and actual Motorola market data on DRAM capacities and prices. Sources: [49, 63].

\multicolumn{7}{c}{ICs: DRAMs}						
1	2	3	4	5	6	7
Year	DRAM (bits)	Die Area (cm^2)	Capacity (Model Results) (MB)	Capacity (Actual Data) (MB)	Price (Model Results) ($)	Price (Actual Data) ($)
1984	16K	0.20	0.005	0.002	29	1.09
1984	64K	0.24	0.008	0.008	31	3.4
1984	256K	0.40	0.043	0.032	42	17.9
1986	1M	0.60	0.13	0.125	62	100
1988	4M	0.80	0.34	0.5	88	264
1991	16M	1.34	1.91	2	396	329

and price per megabyte of DRAMs are provided.

DRAMs Capacities and Prices

Table 4.7 presents the model results and the actual Motorola market data on DRAM capacities and prices. For each DRAM die area, the chip's memory capacity and price are computed, and the results are listed in columns 4 and 6 of the table. The model results on DRAM capacities and the actual Motorola market data in Table 4.7 are almost identical. However, the 1984 model price results and actual data differ significantly because the old technologies were nearing obsolescence. As new technologies are introduced, the demand for old ones diminishes, forcing their prices down. After a while, their market prices are even depressed below a fully marked up production cost.

The price differences in 1986 and 1988 show that either Motorola used rather large markups during their initial pricing of the 1 and 4 megabit DRAMs or they could not achieve high enough die yields to price their DRAMs lower. It has been suggested that the Motorola DRAM prices presented in Table 4.7 left a huge market window for the Japanese ICs manufacturers to infiltrate the DRAM market with prices far be-

Table 4.8: Model results and actual Motorola market data on the price per megabyte of DRAMs. Sources: [49, 63].

ICs: DRAMs		
1	2	3
Year	Price per Megabyte (Model Results) ($)	Price per Megabyte (Actual Data) ($)
1984	1060.5	559.3
1986	476	800
1988	260	528
1991	207	164.5

low Motorola's—a phenomenon known at the time as "market memory dumping and flooding." The Motorola 1 and 4 megabit DRAM prices were cut by more than 50 percent six months after their market introduction in 1986 and 1988 [63], so perhaps the suggestion has some merit.

DRAM Price per Megabyte

Table 4.8 presents the model results and the actual Motorola market data on the DRAM price per megabyte. The model results and the actual market data presented in columns 2 and 3 further illustrate possible Motorola overpricing or low production yields. It was not until 1991 that the actual price per megabyte of DRAM was reduced enough to be consistent with the model results.

DRAM Capacity per cm^2

Table 4.9 presents the model results and the actual Motorola market data on the number of DRAM megabytes per unit die area. The model results and the market data are very similar. The trend reflects a consistent increase in the DRAM density, partly traceable to the trend of decrease in feature size. As the DRAM density increases and the price per megabyte decreases, semiconductor DRAMs might replace the magnetic hard disk as the main computer storage technology, as soon as nonvolatile ones are

Table 4.9: Model results and actual Motorola market data on the number of DRAM megabytes per cm^2. Sources: [49, 63].

ICs: DRAMs		
1	2	3
Year	#Megabytes per cm^2 (Model Results)	#Megabytes per cm^2 (Actual Data)
1980	0.02	n.a.
1981	0.03	0.034
1982	0.05	n.a.
1983	0.07	n.a.
1984	0.11	0.08
1985	0.14	n.a.
1986	0.22	0.21
1987	0.33	n.a.
1988	0.43	0.625
1989	0.65	n.a.
1990	0.96	n.a.
1991	1.43	1.49

developed. DRAMs are faster and more reliable. (In subsection 4.8.1 of this book, the ICs and the magnetic storage supply models are used to illustrate a case in which the price per megabyte of DRAMs becomes cheaper than that of magnetic storage by the turn of the century.)

4.4 Magnetic Storage: Model Results and Actual Market Data

Model results and actual market data on magnetic hard disk price per megabyte and areal densities are compared in subsections 4.4.1 and 4.4.2. Subsection 4.4.3 explains why magnetic hard disk volumetric density and data rate results from the model are not compared to actual market data.

4.4.1 Magnetic Hard Disk Price per Megabyte

For more than 30 years, the price per megabyte of magnetic hard disk has been decreasing. Several factors have contributed to the decrease, including the increase in the manufacturing yields of magnetic hard disks with smaller track pitches and bit cell lengths and the decrease in the prices of fast CISC CPUs and high capacity DRAMs.

Table 4.10 presents in columns 2 and 3 the model results and the actual data on the price per megabyte of magnetic hard disk. The model results and the actual market data differ, on average, by less than 3 percent, and an examination of the trends reveals that the values dropped by almost 90 percent over a period of 11 years.

4.4.2 Magnetic Hard Disk Areal Density

Table 4.11 presents in columns 2 and 3 the model results and the actual market data on the magnetic hard disk areal densities. Most of the model results in the second column of Table 4.11 are slightly larger than the market data in the third column, but by less than a 12 percent margin. Note that the actual areal density of magnetic storage in 1991 is more than 34 times larger than the actual areal density in 1980, while the actual price per megabyte of magnetic storage in 1991 is only about one-ninth that of 1980 (see Table 4.10). This suggests that in the 1980s magnetic hard disk manufacturers were pressed to achieve higher magnetic storage densities to compete with optical storage. Furthermore, they could afford not to keep their prices apace with areal density

Table 4.10: Model results and actual market data on the price per megabyte of magnetic hard disks. Source: [90].

Magnetic Hard Disks		
1	2	3
Year	Price per Megabyte (Model Results) ($)	Price per Megabyte (Actual Data) ($)
1980	56.23	56.23
1981	46.21	46.16
1982	37.52	37.90
1983	30.83	31.11
1984	25.07	25.54
1985	20.60	20.97
1986	16.76	17.21
1987	13.66	14.13
1988	11.22	11.60
1989	9.15	9.52
1990	7.46	7.82
1991	6.09	6.42

Table 4.11: Model results and actual market data on the areal density of magnetic hard disks. Source: [58].

Magnetic Hard Disks		
1	2	3
Year	Areal Density (Model Results) (MB/cm^2)	Areal Density (Actual Data) (MB/cm2)
1980	0.16	0.09
1981	0.21	0.13
1982	0.27	0.18
1983	0.35	0.25
1984	0.46	0.34
1985	0.60	0.46
1986	0.78	0.64
1987	1.01	0.88
1988	1.32	1.20
1989	1.72	1.65
1990	2.24	2.27
1991	2.92	3.12

improvements because magnetic storage response time is much shorter than optical, and its price per megabyte is much less[7].

4.4.3 Notes on Volumetric Density and Data Rate

Volumetric density depends on hard disk design factors like the number of coated disk surfaces and the head-medium setup height, which affects the number of disks per storage device height. Since these factors are manufacturer-dependent, no collective volumetric density market trends are available, and no model results will be presented for comparison.

The hard disk *data rate* is itself dependent on the hard disk rotational speed and the number of disk surfaces per unit storage compo-

---

[7]The suggestion does not exclude the large R&D costs the magnetic storage manufacturers have to pay to achieve the high areal density; however, these costs are not high enough to keep the magnetic storage price per megabyte rate of decrease at approximately one-third the rate of increase of areal density.

Table 4.12: Model results and actual market data on the price per megapixel of a 19-inch color CRT display. Sources: [1, 2, 16, 28, 39, 60, 76, 82, 84, 95].

19-inch Color CRTs		
1	2	3
Year	Price per Megapixel (Model Results) ($/MP)	Price per Megapixel (Actual Data) ($/MP)
1985	8,035	8,000
1986	6,391	6,000
1987	5,110	2,900
1988	4,316	2,380-3,800
1989	3,467	2,400-5,000
1990	2,795	n.a.
1991	2,261	2,300

nent height. Here, too, no general market trend for the data rate was available. Consequently, no comparative model results are provided.

4.5 Color CRT Display: Model Results and Actual Market Data

The 19-inch color CRT display was introduced in 1985 and became the principal workstation and CAD display [16]. Two color CRT mask technologies have dominated the market, the Sony Trinitron and the small holes perforated metal shadow mask technologies. In the display model, only the technology trends for metal shadow masks perforated with small holes were incorporated.

Model results and actual market data on the color CRT price per megapixel and number of pixels per inch are presented in subsections 4.5.1 and 4.5.2, respectively.

4.5.1 Color CRT Price per Megapixel

Table 4.12 presents in columns 2 and 3 the model results and the actual

Table 4.13: Model results and actual market data on the number of pixels per inch of a color CRT display. Sources: [1, 2, 16, 28, 39, 60, 76, 82, 84, 95].

19-inch Color CRTs		
1	2	3
Year	#Pixels per inch (Model Results)	#Pixels per inch (Actual Data)
1985	54	76
1986	59	76
1987	65	84
1988	71	67
1989	78	76-135
1990	85	n.a.
1991	93	84

market data on the price per megapixel of a 19-inch color CRT display. Most of the values in the two columns differ by less than 10 percent for each corresponding year. It is interesting to note that a 19-inch color CRT display in 1985 was priced as much as an average computer workstation in 1991. Very few single users could afford such displays. Nevertheless, with improved CRT and metal shadow mask manufacturing yields, and decreased costs of CISC CPUs and DRAMs, the CRT prices dropped by more than 70 percent in a five-year period. The price drop would have been even greater had there been a strong competing display technology. But the 19-inch color LCD, the best contender, is not yet available to challenge the CRT's reign.

4.5.2 Color CRT Number of Pixels per Inch

In Table 4.13, the model results and the actual market data on the number of pixels per inch are listed in the second and third columns. (Note that the number of pixels displayed on the screen is equivalent to the display resolution.) The actual data range from 76 to 135 pixels per inch, while the display model results were tuned to increase the number of pixels per inch from 54 to 93 over six years, an increase of nearly 7 pixels per inch/per year. The model was not tuned to reach the 135 pixels per inch value because the display manufacturers have no threat of competition in the resolution race, and the 135 pixels per inch value does not reflect the 1989 or 1991 color CRT resolutions available on the market. (Manufacturers can afford to increase the number of pixels per inch at a slower pace than their manufacturing laboratories can achieve.)

4.6 UNIX: Model Results and Actual Market Data

Most activities in UNIX today are performed on porting or enhancing its functionality to make it compatible with all the available hardware platforms. The following section presents the model results and the actual market data on the UNIX porting and development-from-scratch time periods and costs in subsections 4.6.1 and 4.6.2, respectively.

4.6.1 UNIX Porting Times and Costs

Table 4.14 presents the model results and the actual market data on the UNIX porting time periods and costs. Unfortunately, the market data were only available for the years 1980 and 1991. Nevertheless, Table 4.14 presents the model results for the 11-year study period. The results display a decreasing trend, and both the porting time period and the corresponding cost results match the available data within 1 percent. Of course, the model was tuned to match the actual data. The model cost results were not marked up by 200 percent because the development-from-scratch or the porting costs are incurred by the software manufacturer or developer while developing the software. The final product price will depend on the hardware platform on which it will run and on the software's market demand [73].

4.6.2 UNIX Development-from-Scratch Time and Cost

UNIX development-from-scratch is not as prevalent today as it was in the early 1980s, when every UNIX-based workstation manufacturer developed from scratch its own version of UNIX for its hardware platform. UNIX development-from-scratch took about 15 man-years in 1980 and about 10 man-years in 1991. Table 4.15 presents the model results and the actual market data on the UNIX development-from-scratch time periods and costs. Again, the market data were only available for the years

Table 4.14: Model results and actual market data on the UNIX porting time periods and costs. Source: [73].

UNIX: Porting				
1	2	3	4	5
Year	Porting Time (Model Results) (Yrs)	Porting Time (Actual Data) (Yrs)	Porting Cost (Model Results) ($)	Porting Cost (Actual Data) ($)
1980	1.50	1.5	210,000	210,000
1981	1.46	n.a.	198,979	n.a.
1982	1.40	n.a.	185,045	n.a.
1983	1.37	n.a.	175,351	n.a.
1984	1.31	n.a.	163,410	n.a.
1985	1.26	n.a.	151,755	n.a.
1986	1.21	n.a.	141,659	n.a.
1987	1.16	n.a.	132,415	n.a.
1988	1.12	n.a.	123,712	n.a.
1989	1.08	n.a.	115,775	n.a.
1990	1.03	n.a.	107,250	n.a.
1991	0.99	1	100,553	100,000

Table 4.15: Model results and actual market data on the UNIX development-from-scratch time periods and costs. Source: [73].

UNIX: Development-from-Scratch				
1	2	3	4	5
Year	Development Time (Model Results) (Yrs)	Development Time (Actual Data) (Yrs)	Development Cost (Model Results) ($)	Development Cost (Actual Data) ($)
1980	15.00	15	2,100,000	2,100,000
1981	14.64	n.a.	1,989,785	n.a.
1982	14.02	n.a.	1,850,450	n.a.
1983	13.69	n.a.	1,753,510	n.a.
1984	13.14	n.a.	1,634,101	n.a.
1985	12.57	n.a.	1,517,553	n.a.
1986	12.08	n.a.	1,416,595	n.a.
1987	11.63	n.a.	1,324,151	n.a.
1988	11.19	n.a.	1,237,116	n.a.
1989	10.79	n.a.	1,157,750	n.a.
1990	10.30	n.a.	1,072,502	n.a.
1991	9.94	10	1,005,527	1,000,000

1980 and 1991; nevertheless, Table 4.15 presents the model results for the 11-year study period. The results display a decreasing trend, and both the development-from-scratch time period and cost results match the available data within 1 percent. Here, too, the model was tuned to match the actual data.

4.7 Workstation Assembly Model: Inputs and Projected Results

The inputs of the workstation assembly process model are:

1. The study period starts in 1991 and ends in 1996. The year 1991 can be used to compare the model's results with actual workstation prices. Because of rapid technological changes in the industry, a period of five years is sufficiently long, given relatively short product lives and innovation uncertainties.

2. The component supply simulation models presented in earlier sections of this chapter were used to project the price per megaHertz of CPUs, the price per megabyte of DRAMs, the price per megabyte of magnetic hard disk, and the price of a 19-inch color CRT from 1991 to 1996. It is assumed that the component supply models' input parameters and the rates of change of each of the components' physical characteristics trends listed in section 4.2 hold for the 1980–1996 period. These projected results[8] and other assumptions are listed on Table 4.16. The price of the operating system is set at $800 [87], and the prices of each electric power supply, mounting board, cable, mouse, and keyboard are set at $100, $20, $50, $50, and $100, respectively [33]. Since all the components' costs have been marked up by 200 percent in Chapter 4, no additional workstations' price markups are included in the model.

Each workstation has a vector of attributes, and these attributes determine the price of the workstation. For the sake of comparison, three types[9] of workstations are considered, and it is assumed that the manufacturer has the capabilities to produce all of them:

---

[8]The projection years are actually 1992 to 1996 because the 1991 results can be validated with actual market data.

[9]The workstation type refers to the vector of workstation *attributes*: the speed of the processor, the amount of main memory, the capacity of the hard disk, the

4.7. WORKSTATION ASSEMBLY MODEL

Table 4.16: Projected prices of the workstation assembly components and raw materials. Sources: [33, 87].

	Process Model					
	1991	1992	1993	1994	1995	1996
CPU($/MHz)	7.1	6.8	6.5	3.08	2.9	2.7
DRAM($/MB)	79	60	35	25	20	14
Disk($/MB)	6.1	5	4.1	3.3	2.7	2.2
UNIX (each)	800	800	800	800	800	800
Display (each)	3,416	3,322	3,232	3,146	3,064	2,985
Power supply (each)	100	100	100	100	100	100
Board (each)	20	20	20	20	20	20
Cable (each)	50	50	50	50	50	50
Mouse (each)	50	50	50	50	50	50
Keyboard (each)	100	100	100	100	100	100

1. Type 1 (Economy Model): 50 megaHertz CPU, 16 megabytes of DRAM, 250 megabytes of magnetic storage, a 19-inch color CRT monitor, and a UNIX operating system with the documentation.

2. Type 2 (Mid-Range Model): 100 megaHertz CPU, 32 megabytes of DRAM, 500 megabytes of magnetic storage, a 19-inch color CRT monitor, and a UNIX operating system with the documentation.

3. Type 3 (Top-of-the-Line Model): 200 megaHertz CPU, 64 megabytes of DRAM, 1,000 megabytes of magnetic storage, a 19-inch color CRT monitor, and a UNIX operating system with the documentation.

4.7.1 Projected Results

Figure 4.1 presents a log-linear graph that shows how the present value prices of assembled workstation types 1, 2, and 3 decrease over the 1991–1996 period. The objective function of the linear process model [equation (3.93)] was minimized for each of the proposed workstation types, and the

---

resolution of the display, and the software programs loaded into the machine's hard disk.

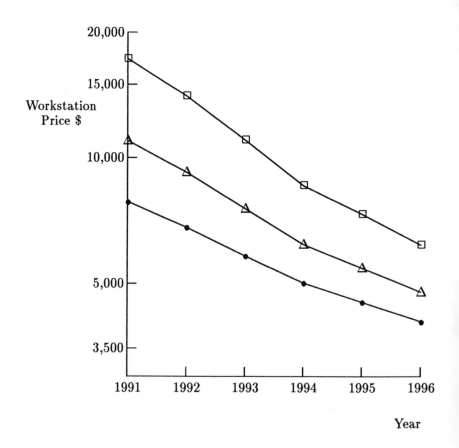

Figure 4.1: Present value prices of workstations: types 1, 2, and 3.

base 10 logarithm of their present value prices are plotted in Figure 4.1 for each year of the 1991–1996 study period.

Figure 4.1 shows that the present value prices of all workstation types decrease by more than 65 percent in five years. This is a rate of decrease of over 20 percent per year for fixed capabilities workstations. (In 1996, the price of a 19-inch color CRT display is close to 60 percent of the total price of an Economy Model workstation—type 1.) These results do not reflect actual workstation prices. Their sole purpose is to show the correlation between the price decrease of the assembled workstation and its components.

The price of the top-of-the-line workstation will be less than $10,000 by 1994. Not everyone will be happy about this. For example, Cray Research[10] sounded a warning in 1991 [59], acknowledging that the most serious threat to its business is the emergence of the superworkstation which, for less than $10,000, will be capable of performing the same computations that a supercomputer[11] does now—and in only twice the time!

Remark

Since the workstation assembly model did not include plant expansion costs, or import, export, and government quota restrictions, and it considered only one assembly plant and one market in its vicinity—that is, no transportation costs were involved—its formulation can forego cost minimization in the objective function [equation (3.93)]. The costs of the assembled workstations can be computed by multiplying the configuration requirements of each workstation type by the corresponding component prices in Table 4.16. Nevertheless, the presented assembly model was formulated as such to provide a foundation upon which future research can expand.

4.8 Sensitivity Analyses

To mainframe and supercomputer manufacturers, the emergence of the workstation has meant cuts in market shares and profit margins [59]. In response, mainframe manufacturers have transferred their sophisticated

---

[10] Cray Research is the manufacturer of the fastest supercomputers in the world, to date.

[11] A typical supercomputer costs $2 to $3 million.

technologies to the workstation platform and introduced high end workstations which can outperform their original mainframes. HP announced in 1991 a high-end workstation based on its mainframe's RISC Precision Architecture platform, and IBM announced in 1990 a high-end workstation which is based on its mainframe's superscalar RISC architecture. All of these technology migrations have become possible because manufacturers have achieved high production yields as the technology-driving trends of their hardware improve.

What will shape the trends of the future? This is impossible to answer with certainty. We think, however, that models of the type developed in this book can help us understand where the industry might be going. To illustrate, this section presents sensitivity analyses of how the attributes of assembled workstations might change in response to changes in certain technology-driving trends. The discrete event simulation supply models and the linear workstation assembly process model provided in Chapter 3 of this book are used to perform these analyses. Each change in a technology-driving trend is represented with a particular case number. In turn, the results of each change case are compared with the base case results for the period of analysis.

This section examines the models' behavior in response to the following two "What if...?" questions:

1. How sensitive are the prices of workstation components and assembled workstations to a faster rate of decrease and to no decrease in the IC's *feature size* (FS)?

2. How sensitive are the IC die yields and the CPU and DRAM prices to an increase and to a decrease in the *number of silicon wafer defects per unit area* (DPUA)?

The results are presented in the following subsections. Subsection 4.8.1 studies the effects of the feature size (FS) of ICs on the prices of the components and, ultimately, on the prices of the assembled workstations. The special case wherein the price per megabyte of DRAM becomes cheaper than the magnetic hard disk's is also discussed. (The effects of the feature size on the pricing of one copy of the UNIX operating system are not presented in this section since only the UNIX development-from-scratch and porting costs are modeled in section 3.5. In general, the price of the operating system is equal to a fixed markup to the overall workstation cost, set in this book at $800 [87].)

4.8. SENSITIVITY ANALYSES

Subsection 4.8.2 studies the effects of the silicon wafer defects per unit area (DPUA) on the die yields and, consequently, on the price per megaHertz of CPUs and the price per megabyte of DRAMs.

4.8.1 Sensitivity to the Feature Size

The feature size is one of the main technology-driving trends of the VLSI technologies. It has been estimated that the feature size will not run into any physical barriers until the end of the century due to the recent achieved improvements in lithography and silicon etching processes [21].

However, there is concern among many in the electronics industry that the feature size might reach a physical barrier and stop decreasing[12] in the next decade [18, 21]. To study the effects of such an event, the models developed in Chapter 3 were simulated with a stoppage in decreasing feature size after 1992. The year 1992 was chosen instead of the year 2000, for example, because of the rapid pace at which the computer industry is changing and the uncertainties associated with it; a later year may be too far into the future for us to analyze the consequences of a stoppage in decreasing feature size. Again, feature size sensitivity analyses are meant to shed some light on how important a continually decreasing feature size is to the electronics industry and the computer industry in general. To make the study complete, the models were also simulated with an accelerated decrease of the feature size after 1992.

The alternative feature size trends were implemented as follows:

- **Case 1 — Base Case**
  The feature size follows the trend presented in equation (3.14):

$$FS_t = FS_{BC} = 10^{1.4-0.055*(t-1960)} \ (micron). \qquad (4.3)$$

  In this case the feature size continually decreases at 5.5 percent per year for the period of study.

- **Case 2 — $FS_t = FS_{BC/1992}$ for t > 1992**
  The feature size stops decreasing after 1992; hence, for $t > 1992$,

---

[12]A physical barrier might only slow down the decrease in the feature size. Since this is a "What if...?" illustration and there is no way to predict with certainty the consequences of the physical barrier, only the extreme case—stoppage of feature size decrease—was considered.

the feature size values are equal to the base case's value in 1992:

$$FS_t = FS_{BC/1992} = 10^{1.4-0.055*(1992-1960)} \ (micron) \ \text{for} \ t > 1992. \tag{4.4}$$

- **Case 3 — $FS_t < FS_{BC}$ for t > 1992**
  The feature size decreases faster than expected after 1992; hence, for $t > 1992$, the feature size values are equal to the base case's value in 1992 multiplied by the new rate of decrease of 10 percent:

$$FS_t = FS_{BC/1992} * 10^{-0.1*(t-1992)} \ (micron) \ \text{for} \ t > 1992. \tag{4.5}$$

For each of the cases presented above, the ICs, magnetic hard disk, and color CRT display supply models of Chapter 3 were simulated for the 1992–1996 period. It is assumed that the component supply models' input parameters and the rates of change of each of the components' physical characteristics trends listed in section 4.2 hold for the 1980–1996 period.

The organization of this subsection is as follows:

- The first four paragraphs illustrate the sensitivity to changes in the feature size of the price per megaHertz of CPUs, the price per megabyte of DRAMs, the price per megabyte of magnetic hard disks, and the prices of a 19-inch color CRT for each of the three cases presented above.

- The fifth paragraph illustrates the collective effect of the changes in the feature size on the assembled workstation prices for each of the three cases presented above.

- The sixth paragraph illustrates how, if feature size decreases at a faster rate than the base case's, the price per megabyte of DRAMs will become lower than the price per megabyte of magnetic storage in the early part of the next decade.

Sensitivity of the CPU Price per MegaHertz to Feature Size

Figure 4.2 illustrates in a log-linear graph the decreasing price per megaHertz of CPUs for the 1992–1996 period. The ICs supply model was simulated to compute for each of the three cases the prices of CPUs with three different speeds—50, 100, and 200 megaHertz. The prices of

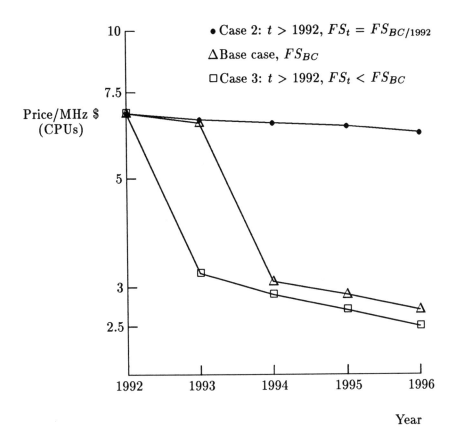

Figure 4.2: Feature size: sensitivity of the price per megaHertz of CPUs — cases 1, 2, and 3.

the CPUs were then divided by their corresponding speeds to compute the price per megaHertz values. The base 10 logarithm of the lowest price per megaHertz values are plotted in Figure 4.2 for each year of the 1992–1996 period.

It is evident that, when the feature size stops decreasing, the price per megaHertz values are higher than the base case's. Furthermore, the price per megaHertz values are lower than the base case values where there exists an accelerated decrease in feature size. By almost doubling the rate of decrease of the feature size, from 5.5 percent to 10 percent, and by stopping its decrease after 1992, the price per megaHertz values differ from the base case values by as much as 7.4 percent and 129.6 percent, respectively, in 1996. However, the most noticeable characteristic of Figure 4.2 is the drop in the price per megaHertz in the base case and case 3 values after 1993 and 1992, respectively. The reason for those two drops is the packaging costs of the CPUs. As the feature size decreases, the die area of a fixed speed CPU gets smaller, and as the die area gets smaller than 1.1 cm^2, the packaging cost drops by \$47 [refer to equations (3.34) and (3.35)]. Hence, the drops in the graphs of Figure 4.2. The case 3 graph drops earlier because the feature size is decreasing faster than in the base case.

Sensitivity of the DRAM Price per Megabyte to Feature Size

Figure 4.3 illustrates in a log-linear graph the decreasing price per megabyte of DRAMs for the 1992–1996 period. The ICs supply model was simulated to compute, for each of the three cases, the prices of DRAMs with six different capacities—1, 2, 8, 32, 64, and 128 megabytes. The prices of the DRAMs were then divided by their corresponding capacities to compute the price per megabyte values. The base 10 logarithm of the lowest DRAM price per megabyte values are plotted in Figure 4.3 for each year of the 1992–1996 period.

It can be seen in Figure 4.3 that, when the feature size stops decreasing, the DRAM price per megabyte values are higher than the base case's. Moreover, the DRAM price per megabyte values are lower than the base case values for an accelerated decrease in feature size. By almost doubling the rate of decrease of the feature size, from 5.5 percent to 10 percent, and by stopping its decrease after 1992, the price per megabyte values of DRAMs differ from the base case values by as much as 55 percent and 114.3 percent, respectively, in 1996. In response to the drop in

4.8. SENSITIVITY ANALYSES

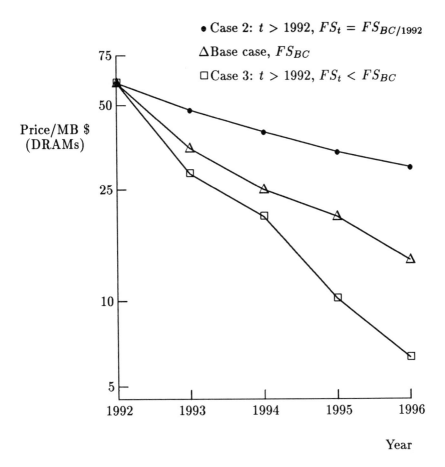

Figure 4.3: Feature size: sensitivity of the price per megabyte of DRAMs — cases 1, 2, and 3.

the packaging costs, the drops in the price per megabyte in Figure 4.3 are not as pronounced as those in Figure 4.2. Nonetheless, they do occur in 1995 and 1994 for the base case and case 3, respectively.

Sensitivity of the Magnetic Hard Disk Price per Megabyte to Feature Size

Figure 4.4 illustrates in a log-linear graph the decreasing price per megabyte of magnetic hard disks for the 1992–1996 period. The magnetic hard disk supply model was simulated to compute for each of the three cases the hard disk price per megabyte values. The feature size affects the performance of the channel's microcontroller [equation (3.65)], which in turn affects the price per megabyte values of magnetic hard disks [equation (3.62)]; refer to the magnetic hard disk supply model of section 3.3 for more details.

It can be seen in Figure 4.4 that when the feature size stops decreasing, the magnetic hard disk price per megabyte values are higher than the base case's. The figure further shows that the price per megabyte values are lower than the base case values when there exists an accelerated decrease in feature size. By almost doubling the rate of decrease of the feature size, from 5.5 percent to 10 percent, and by stopping its decrease after 1992, the price per megabyte values of magnetic hard disks differ from the base case values by as much as 9.1 percent and 18.2 percent, respectively, in 1996.

Sensitivity of the Color CRT Prices to Feature Size

Figure 4.5 illustrates in a log-linear graph the decreasing prices of a 19-inch color CRT display for the 1992–1996 period. The CRT display supply model was simulated to compute for each of the three cases the 19-inch color CRT prices. The feature size affects the speed and the capacity of the microprocessors and the DRAMs used in the CRT image drivers. Consequently, the feature size affects the number of pixels per inch displayed on the screen [equation (3.77)], the screen's resolution [equation (3.71)], and the cost per megapixel [equation (3.81)].

It can be seen in Figure 4.5 that, when the feature size stops decreasing, the 19-inch color CRT display prices are higher than the base case's. Moreover, the 19-inch color CRT prices are lower than the base case prices where there exists an accelerated decrease in feature size. By

4.8. SENSITIVITY ANALYSES

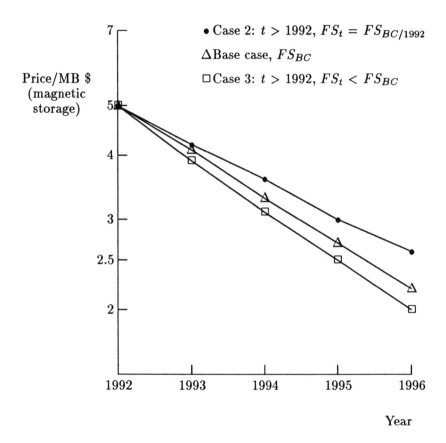

Figure 4.4: Feature size: sensitivity of the price per megabyte of magnetic hard disks — cases 1, 2, and 3.

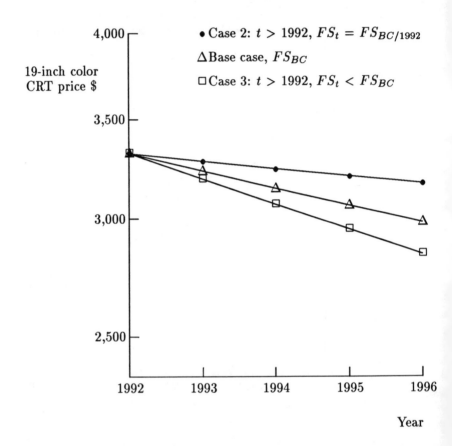

Figure 4.5: Feature size: sensitivity of the price of a 19-inch color CRT display — cases 1, 2, and 3.

almost doubling the rate of decrease of the feature size, from 5.5 percent to 10 percent, and by stopping its decrease after 1992, the 19-inch color CRT prices differ from the base case prices by as much as 4.8 percent and 6.3 percent, respectively, in 1996.

Sensitivity of the Assembled Workstations Prices to Feature Size

As has been shown, the feature size affects the attributes and prices of the workstation hardware components. In turn, it also affects the price of the workstation. The workstation assembly process model in section 3.6 was used to illustrate the effects of changes in the feature size on the price of an assembled workstation with a 100-megaHertz CPU set, a 32-megabyte main memory, a 500-megabyte magnetic hard disk, and a 19-inch color CRT. Figure 4.6 presents the magnitude of the present value of the workstation prices for the three feature size cases.

It can be seen that when the feature size stops decreasing, the workstation prices are higher than the base case prices. The figure also shows that the workstation prices are lower than the base case prices for a faster decreasing feature size. By almost doubling the rate of decrease of the feature size, from 5.5 percent to 10 percent, and by stopping its decrease after 1992, the workstation prices differ from the base case prices by as much as 8.4 percent and 20.5 percent, respectively, in 1996.

Feature Size: Sensitivity of the Price per Megabyte — DRAMs versus Magnetic Storage

One of the main objectives of the Sematech consortium was to develop denser DRAM ICs and help the United States regain some of its lost DRAM market share from the Japanese manufacturers. But the good news for the U.S. ICs manufacturers is a source of worry for the magnetic hard disk manufacturers. Not only do they need to fend off the competition from the optical technology but denser DRAMs might have a lower price per megabyte, and if nonvolatile[13] ones are developed, DRAMs might replace the magnetic hard disk as the main computer storage component.

---

[13] A nonvolatile DRAM retains the data stored in it even when the power is switched off.

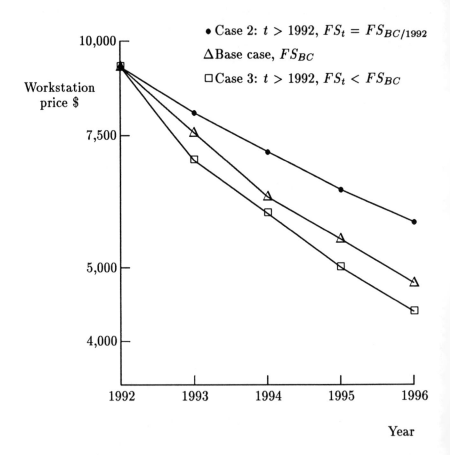

Figure 4.6: Feature size: sensitivity of the present value price of a type 2 workstation — cases 1, 2, and 3.

4.8. SENSITIVITY ANALYSES

In this paragraph, a scenario is analyzed that results in the DRAM price per megabyte becoming cheaper than magnetic storage. If the ICs and magnetic results presented in Chapter 4 are projected to the year 2000, the DRAM price per megabyte will not catch up with magnetic storage during this decade. However, if the feature size decreases at a faster rate (case 3) than was projected in the base case, and the magnetic storage manufacturers are not able to improve their products' attributes at a faster rate than was presented in section 3.3, the DRAM price per megabyte might become cheaper than magnetic storage during the first half of the next decade. What follows is an illustration of this phenomenon.

Only the base case and case 3 of the feature size trends are considered for this analysis. For each one of the cases, the ICs and magnetic storage supply models of Chapter 3 were simulated for the 1992–2005 period and the results are presented in Figure 4.7. It is assumed that the ICs and magnetic storage supply models' input parameters and the rates of change of each of the components' physical characteristics trends listed in section 4.2 hold for the 1980–2005 period.

Figure 4.7 illustrates in a log-linear graph the decreasing prices per megabyte of DRAMs and magnetic storage for the 1992–2005 period. The ICs and magnetic storage supply models were simulated to compute for each of the two cases the prices of DRAMs with seven different capacities—1, 2, 8, 32, 64, 128, and 256 megabytes—and the price per megabyte of magnetic storage. The prices of the DRAMs were then divided by their corresponding capacities to compute the price per megabyte values. The base 10 logarithm of the lowest DRAM price per megabyte values are plotted in the upper two graphs of Figure 4.7 for each year of the 1992–2005 period, and the magnetic storage price per megabyte values are plotted in the lower two graphs for each of the two cases. As can be seen in Figure 4.7, the DRAM price per megabyte becomes less than the magnetic storage price per megabyte around 2001.

The fact that the magnetic storage price per megabyte values do not decrease faster than projected can be attributed to the following. As the feature size decreases faster than expected, the microcontroller in the hard disk's channel can operate faster [see equation (3.22)], the price per microcontroller MIPS decreases (see section 4.3), and, consequently, the price per megabyte values of magnetic storage become less than the base case's [see equations (3.62) and (3.65)]. However, the channel's performance improvement does not affect the price per megabyte of magnetic

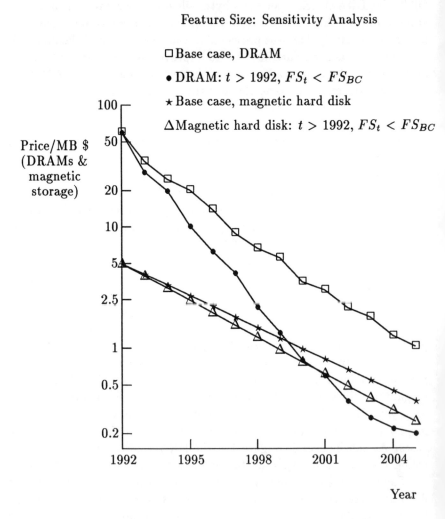

Figure 4.7: Feature Size: Sensitivity of the DRAM and the magnetic hard disk prices per megabyte — cases 1 and 3.

storage enough to decrease it faster than the DRAM's. This is because the net decrease of the DRAM price per megabyte values is approximately 2 over the period of study [see equations (3.25) and (3.9)], while the net decrease in value of the magnetic storage price per megabyte values is only about 0.3 [see equations (3.22), (3.65), and (3.9)], yielding, approximately, a 7:1 ratio of difference in their sensitivity to the feature size.

4.8.2 Sensitivity to the Number of Silicon Wafer Defects per Unit Area

The die yield plays a crucial role in determining the price of an IC and whether the IC ought to remain in the production line or be removed. As the production process of ICs gets more and more complicated, precision and defects-free production become harder to achieve. One of the main steps in producing defect-free and cheaper ICs is to reduce the number of silicon wafer defects per unit area (DPUA). A defect can occur anytime during the production process of the wafer itself or the IC.

Clean room technologies and learning have been two key factors in improving the yields of good manufactured dies. Since DPUA reflects how clean, efficient, and accurate the manufacturing of an ICs plant is, its effect on the ICs prices will be studied in this subsection, in particular, the price per megaHertz of CPUs and the price per megabyte of DRAMs. DPUA was assumed to remain constant at 2.5 defects per cm^2 during the 1980–1991 simulation period. To study the effects of varying DPUA, the ICs supply model was simulated from 1992 to 1996, with DPUA set in one case to twice its original value—5 defects per cm^2—and in a second case to half its original value—1.25 defects per cm^2—after 1992. It is assumed that the ICs supply models' input parameters and the rates of change of each of the ICs' physical characteristics trends listed in section 4.2 hold for the 1980–1996 period.

What follows are the three cases used to show the effects of changes in the number of silicon wafer defects per unit area on the IC die yields, the price per megaHertz of CPUs, and the price per megabyte of DRAMs:

- Case 1 — Base Case

$$DPUA_t = DPUA_{BC} = 2.5 \ defects/cm^2 \ \text{for} \ t > 1992. \quad (4.6)$$

Table 4.17: DPUA: sensitivity of the die yields for certain die areas — cases 1, 2, and 3.

	Sensitivity Analysis		
	Case 2	Base Case	Case 3
Die Area (cm^2)	Die Yield (%) DPUA = 5	Die Yield (%) DPUA = 2.5	Die Yield (%) DPUA = 1.25
0.0625	66.8	77.3	88.3
0.2601	30.2	49.8	66.1
0.5776	11.5	27.3	47
1.0404	4.1	13.4	30.2
1.6129	1.6	6.6	18.8
2.3104	0.7	3.3	11.5
3.1684	0.3	1.6	6.8
4.1209	0.1	0.9	4.1

- **Case 2**

$$DPUA_t = 2 * DPUA_{BC} = 5 \ defects/cm^2 \ \text{for } t > 1992. \quad (4.7)$$

- **Case 3**

$$DPUA_t = \frac{1}{2} * DPUA_{BC} = 1.25 \ defects/cm^2 \ \text{for } t > 1992. \quad (4.8)$$

DPUA: Sensitivity of the IC Die Yields

The first factor reflecting the change in the value of $DPUA_{BC}$ is the die yield. Equation (4.9) shows that the higher (or the lower) the DPUA number is, the lower (or the higher) the manufacturing die yield:

$$DY_t = WY_t * \left(1 + \frac{DPUA_t * DA_t}{CM_t}\right)^{-CM_t}. \quad (4.9)$$

Table 4.17 lists in the columns labeled case 2, base case[14], and case 3 the die yield model results for these cases. When the $DPUA_{BC}$ increases by 100 percent to 5 defects per cm^2, the die yield values in column 2

---
[14]The base case die yield model results are taken from Table 4.1.

4.8. SENSITIVITY ANALYSES

decrease by as much as 88.9 percent. Similarly, as $DPUA_{BC}$ decreases by 50 percent to 1.25 defects per cm^2, the die yield values in column 4 increase by as much as 355.6 percent, or almost seven times the decrease in the base case DPUA value. The die yield results for the three cases reflect how sensitive the die yield is to clean and precise production environments and processes. A 10 percent decrease in the DPUA could result in a 70 percent improvement in the die yield and, consequently, a sizable decrease in the die prices over time.

DPUA: Sensitivity of the CPU Price per MegaHertz

Figure 4.8 illustrates in a log-linear graph the decreasing price per megaHertz of CPUs for the 1992–1996 period. The ICs supply model was simulated to compute for each of the three cases the prices of CPUs with three different speeds—50, 100, and 200 megaHertz; the prices of the CPUs were then divided by their corresponding speeds to compute the price per megaHertz values. The base 10 logarithm of the lowest price per megaHertz values are plotted in Figure 4.8 for each year of the 1992–1996 period.

It can be seen in Figure 4.8 that, when the $DPUA_{BC}$ is doubled to five defects per cm^2, the price per megaHertz value jumps to more than double the base case value in 1993, reaching a 277 percent difference in 1996. Moreover, as the $DPUA_{BC}$ is halved to 1.25 defects per cm^2, the price per megaHertz value drops to less than half the base case value in 1993, reaching a 51.8 percent difference in 1996. (Note also the effects of the change in packaging costs in the three graph drops in 1993, which were discussed earlier.)

DPUA: Sensitivity of the DRAM Price per Megabyte

Figure 4.9 illustrates in a log-linear graph the decreasing price per megabyte of DRAMs for the 1992–1996 period. The ICs supply model was simulated to compute for each of the three cases the prices of DRAMs with six different capacities—1, 2, 8, 32, 64, and 128 megabytes; the prices of the DRAMs were then divided by their corresponding capacities to compute the price per megabyte values. The base 10 logarithm of the lowest DRAM price per megabyte values are plotted in Figure 4.9 for each year of the 1992–1996 period.

Figure 4.9 shows that, as the $DPUA_{BC}$ is doubled to five defects per

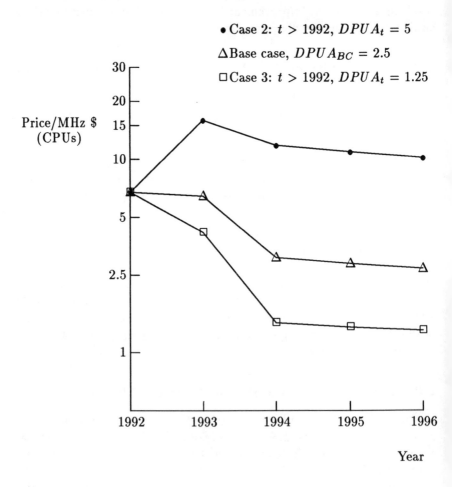

Figure 4.8: DPUA: sensitivity of the price per megaHertz of CPUs — cases 1, 2, and 3.

4.8. SENSITIVITY ANALYSES

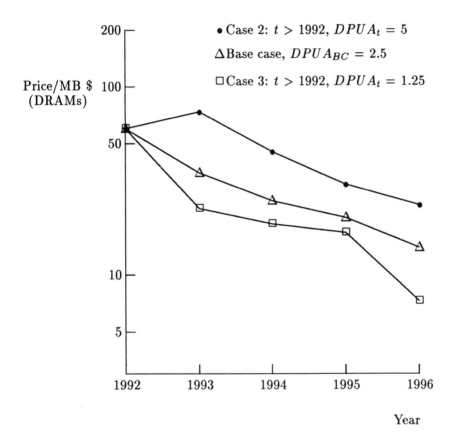

Figure 4.9: DPUA: sensitivity of the price per megabyte of DRAMs — cases 1, 2, and 3.

cm^2, the price per megabyte value jumps to more than double the base case value in 1993, climbing up to a 69.1 percent difference in 1996. When the DPUA$_{BC}$ is halved to 1.25 defects per cm^2, the price per megabyte value drops to less than half the base case value in 1993, reaching a 47.7 percent difference in 1996. (Here, too, note the effects of the difference in packaging costs in the three graph drops in 1993.)

As seen in the percentage differences between the base case and the case 2 and case 3 results, the ICs die yields, the CPU price per megaHertz, and the DRAM price per megabyte results are far more sensitive to an increase in the number of silicon wafer defects per unit area than they are to a decrease.

5

Conclusions and Suggestions for Future Research

This book approached the dynamics of the computer industry through the workstation sector and its components. Discrete event simulation supply models for workstation components, including microprocessors, DRAMs, magnetic hard disks, color CRT displays, and UNIX operating systems were developed to study the effects of improving technology-driving trends on the capabilities and prices of the components and, ultimately, on the attributes of the assembled workstations. A relational diagram concept was used in Chapter 3 to communicate the model relationships of the attributes of the components over time. The supply models were simulated over the 1980–1991 study period to tune their results to match with actual market data. Since the supply models provided *cost* results only, a 200 percent markup was applied to the cost results to compare them with the actual market *price* data. Most of the results of the supply models and the actual market data differed by less than 10 percent.

To study the overall effects of decreasing prices and improving capabilities of the components on the supply of assembled workstations, a linear workstation assembly process model was developed. The objective function was to minimize the present value of the total components and raw materials costs that went into the assembly of the workstations. The price results of the component supply models were projected over the 1992–1996 period and used as input prices in the assembly model. Three workstation types were considered, and all the results showed a

188 CHAPTER 5. CONCLUSIONS AND SUGGESTIONS FOR FUTURE RESEARCH

decreasing trend of workstation prices in the years to come. By 1994, a top-of-the-line workstation will cost less than $10,000 and will have the following hardware capabilities: a 200 megaHertz CPU, a 64 megabyte main memory, a 1 gigabyte magnetic hard disk, and a 19-inch color CRT display with a 2.6 megapixel resolution. These same capabilities, if configured into a workstation in 1991, would cost approximately $20,000. This is a rate of decrease in price of over 20 percent per year for a fixed capabilities workstation.

Finally, sensitivity analyses were performed for the models for three different rates of decrease of the IC feature size and for three different numbers of silicon wafer defects per unit area. It was apparent that all the results of the supply models were sensitive to the changes in the feature size, and their sensitivity was echoed in the overall prices of the assembled workstations. A steady decrease in the feature size helps IC manufacturers stay competitive in a semiconductor market that has become more capital intensive and more competitive than ever. The most striking results of the feature size sensitivity analyses were the prices per megabyte of DRAMs and magnetic storage. As the feature size's rate of decrease doubled, the rates of decrease of both DRAM price per megabyte and magnetic storage price per megabyte increased; however, DRAM price per megabyte decreased at a faster rate than did magnetic storage and was thus projected to catch up with magnetic storage by the year 2001. This suggests that by 2001, if nonvolatile DRAMs were developed, DRAMs might replace the magnetic hard disk as the permanent storage component of computers and change completely the computer system configuration as we know it today.

The results of the ICs supply model showed the most sensitivity to the changes in the number of silicon wafer defects per unit area. As expected, a decrease in the number of silicon wafer defects per unit area during the production of ICs increased the die yields and, ultimately, decreased the prices of DRAMs and CPUs. Since improved IC production learning and clean IC production environments can decrease the number of defects per unit area, efforts to make these improvements can lower the IC manufacturers' direct costs and, ultimately, increase the demand for workstations by allowing lower price-to-performance ratios.

5.1 Suggestions for Future Research

One possible extension to this book would develop supply models of ICs to complement the super high-speed network communication. For instance, gallium arsenide, superconducting, or optical ICs supply models could be developed to study their effects on the performance and prices of future workstations and networks.

A second extension to this research could develop detailed supply and assembly models for liquid crystal displays and optical storage disks. Each of these components has several subcomponents and goes through several manufacturing stages before the final product is assembled. At each stage, there are several manufacturing procedures which must be performed, each with its own yield. Since these technologies are relatively new on the market, their production yield data are kept confidential by the manufacturers. Nevertheless, they will most likely replace the current market dominant CRT display and magnetic storage technologies as soon as their manufacturers gain more experience in producing them; that is, increase their yields, improve their response times, and reduce their costs.

Another extension to this book might study the computer enhancements and the new applications that could result from replacing permanent magnetic storage with a permanent semiconductor one. DRAMs significantly affect the computer system's response time and performance. Moreover, they are simpler and more reliable semiconductor storage elements than their magnetic counterparts, and decreasing their cost per megabyte and improving their density could pave the way, for instance, to real-time computing. Real-time computing requires the computer job to finish within a time constraint or the job is canceled. If DRAMs were used as main memory and as permanent storage, the computer system's access to storage could be sped up by as much as 1 million times (20 milliseconds to 20 nanoseconds), making any storage-intensive application look like a real-time application to the end user. Other new applications will be left for future research to study.

Since most of the computer technologies on the market carry with them uncertainties about their time to obsolescence and about the attributes of their successor technologies and the time of their introduction, further research could introduce uncertainty factors in the supply models and render their behavior stochastic.

A closely related project would study the life cycle of each introduced computer product by subjecting it to competitive forces from more innovative technologies emerging in the market place.

Further research launched from this book could be performed at the company level, where sensitivity analyses might be done on the effects of new technologies introduced on the market. For instance, a sensitivity analysis might be designed to study the effects of RISC technology as the basis of all the processors and microcontrollers used in a workstation.

A linear workstation assembly process model was presented in this book. The size of the model was not prohibitive from a computation time point of view (nevertheless, the process model can increase in size very quickly [47]). Another extension to this research would collect the data and include the following features in a future workstation assembly model: the major workstation manufacturers such as Sun, HP, IBM, DEC, Silicon Graphics, Toshiba, and Motorola; a list of regional markets corresponding to cities worldwide and associated transportation costs; purchases of components and raw materials from around the world with corresponding transportation costs; capacity expansion and depreciation factors of new plant and equipment, taking into consideration the expansion's investment constraints; a market share and life-cycle analyses of old versus new workstation models, focusing especially on the time interval when the prices of the new models start to decrease. These features might be particularly useful to workstation manufacturers to help them estimate where they stand in their production cycles of specific models and to help them define the workstation attributes they need to integrate to remain competitive in such a dynamic industry.

A further extension to this book would study the effects of economies of scale on the supply of computer components and assembled workstations.

Since this book is concerned with supply-side issues, a final extension to this research would study the demand side of computer components and assembled workstations.

5.2 Final Comments

> People and organizations resist change.... Shortage of funds may slow the growth progress, or the inability to increase the market enough to generate the necessary funds to develop technologies, or new applications do not get discovered.... In nature, exponential growth curves always top out. They turn

5.2. FINAL COMMENTS 191

into S-shaped curves...the rapid, exponential expansion of the computer industry will slow down when it outstrips its resources, or whenever the market fails to supply exponentially increasing resources.

This quotation by Myers [65] summarizes the state of the computer industry, the most dynamic industry of our time. By merging the concepts of cost and technical change, this book captures the computer industry's dynamic behavior and sets the stage for future research to be performed in this new field.

References

[1] Apiki, S., and S. Diehl. 1989. "Upscale Monitors." **Byte**, March, pp. 162–174.

[2] Arcuri, F. 1989. "Market Survey: CRT Computer Monitors." **Information Display**, June, pp. 10–13.

[3] Bajorek, C.H. 1989. "Trends in Recording and Control and Evolution of Subsystem Architectures for Data Storage." **IEEE Proceedings, COMPEURO'89: VLSI and Computer Peripherals**, pp. 1-(1–6).

[4] Bell, G. 1988. "Toward a History of Personal Workstations." In **A History of Personal Workstations**, A. Goldberg (ed.). Reading, MA: Addison-Wesley Publishing Company.

[5] Berghof, W. 1989. "Head and Media Requirements for High-Density Recording." **IEEE Proceedings, COMPEURO'89: VLSI and Computer Peripherals**, pp. 1-(97–99).

[6] Bertsekas, D.P., and J.N. Tsitsiklis. 1989. **Parallel and Distributed Computation: Numerical Methods**. Englewood Cliffs, NJ: Prentice Hall, Inc.

[7] Boschma, B.D., et al. 1989. "A 30 MIPS VLSI CPU." **IEEE International Solid-State Circuits Conference: Digest of Technical Papers**, pp. 82–83.

[8] Bowater, R.J. 1989. "The IBM Image Adapter/A^{TM}." **IEEE Proceedings, COMPEURO'89: VLSI and Computer Peripherals**, pp. 1-(104–108).

[9] Breen, P.T. 1988. "Workstations-New Environment, New Market." **Information Display**, November, pp. 8-10.

[10] Brodsky, M.H. 1990. "Progress in Gallium Arsenide Semiconductors." **Scientific American**, February, pp. 68-75.

[11] Brooke, A., D. Kendrick, and A. Meeraus. 1988. **GAMS: A User's Guide**. Redwood City, CA: The Scientific Press.

[12] Cates, R. 1990. "Gallium Arsenide Finds a New Niche." **IEEE Spectrum**, April, pp. 25-28.

[13] Chang, I.F. 1980. "Recent Advances in Display Technologies." **Proceedings of the Society of Information Display**, $\underline{21}$, pp. 45-54.

[14] Chesters, M.J. 1989. "A 1 micron CMOS 128 MHz Video Serializer, Palette and Digital-to-Analogue (DAC) Chip." **IEEE Proceedings, COMPEURO'89: VLSI and Computer Peripherals**, pp. 1-117.

[15] Chow, P. 1991. "RISC (Reduced Instruction Set Computing)." **IEEE Potentials**, October, pp. 28-31.

[16] "Color CRT Prices." 1991. Provided by David Eccles, **Sony Corporation — Engineering Division**, San Diego, CA.

[17] "Computer Confusion: A Jumble of Competing, Conflicting Standards is Chilling the Market." 1991. **Business Week**, June 10, pp. 72-78.

[18] Conversations with Al Tasch, Professor at the Department of Electrical and Computer Engineering, the University of Texas at Austin, Austin, Texas, 1991.

[19] Cunningham, J.A. 1990. "The Use and Evaluation of Yield Models in Integrated Circuit Manufacturing." **IEEE Transactions on Semiconductor Manufacturing**, $\underline{3}$(5), pp. 60-71.

[20] Dill, F.H. 1982. "Future Trends and Mutual Impact of VLSI and Display." **1982 International Display Research Conference**, pp. 9-10.

[21] Flores, G.E., and B. Kirkpatrick. 1991. "Optical Lithography Stalls X-Rays." **IEEE Spectrum**, October, pp. 24–27.

[22] Forsyth, M., S. Mangelsdorf, E. DeLano, et al. 1991. "CMOS PA-RISC Processor for a New Family of Workstations." **COMPCON Spring '91: Digest of Papers**, pp. 202–207.

[23] Funk, H.L. 1989. "Information Displays — An Overview and Trends." **IEEE Proceedings, COMPEURO'89: VLSI and Computer Peripherals**, pp. 2-(1–7).

[24] "Gallium Arsenide Chips: Half Way to Paradise." 1991. **The Economist**, June 15, p. 83.

[25] Garey, M.R., and D.S. Johnson. 1979. <u>Computers and Intractability: A Guide to the Theory of NP-Completeness</u>. New York: W.H. Freeman and Company.

[26] George, J. 1989. "High-Technology Competition between U.S. and Japanese Companies." **Japanese Business Study Program**.

[27] Ghausi, M.S. 1985. <u>Electronic Devices and Circuits: Discrete and Integrated</u>. New York: Holt, Rinehart and Winston.

[28] Goede, W.F. 1982. "Technologies for High-Resolution Color Display." **1982 International Display Research Conference**, pp. 60–62.

[29] Goldberg, A. 1988. <u>A History of Personal Workstations</u>. Reading, MA: Addison-Wesley Publishing Company.

[30] Haemer, J.S., S.P. McCarron, and P.H. Salus. 1991. "Trends in UNIX Software." **COMPCON Spring '91: Digest of Papers**, pp. 365–368.

[31] Hamacher, V.C., Z.G. Vranesic, and S.G. Zaky. 1984. <u>Computer Organization</u>. New York: McGraw-Hill Book Company.

[32] Hayes, F. 1991. "In Search of Standard Unix." **UnixWorld**, December, pp. 91–92.

[33] Hennessy, J.L., and D.A. Patterson. 1989. <u>Computer Architecture: A Quantitative Approach</u>. San Mateo, CA: Morgan Daufman Publishers, Inc.

[34] Hennessy, J.L., and N.P. Jouppi. 1991. "Computer Technology and Architecture: An Evolving Interaction." **Computer**, September, pp. 18-29.

[35] Herold, E.W. 1974. "History and Development of the Color Picture Tube." **Proceedings SID**, 15(4), pp. 141-149.

[36] Hillier, F.S., and G.J. Lieberman. 1986. **Introduction to Operations Research**. Oakland, CA: Holden-Day, Inc.

[37] Hönig, H.E. 1989. "Superconductivity." **IEEE Proceedings, COMPEURO'89: VLSI and Computer Peripherals**, pp. 5-(79-81).

[38] "HP Workstations Performance Data." 1991. Provided by Bryan Brademan, **Hewlett Packard — Sales and Marketing Division**.

[39] Iki, T., and K. Werner. 1989. "CRTs." **Information Display**, December, pp. 6-7.

[40] Infante, C. 1988. "Advances in CRT Displays." **1988 International Display Research Conference**, pp. 9-11.

[41] Infante, C. 1986. "CRT Technology: Progress and Issues." **Proceedings of the Society of Information Display**, 27(4), pp. 245-248.

[42] "Intel's 80x86 Data." 1991. Provided by Todd J. Derr, University of Pittsburgh, Pittsburgh, Pennsylvania.

[43] "Intel's Plan for Staying on Top." 1989. **Fortune**, March 27.

[44] Jain, R. 1991. "Performance Analysis Ratholes or How to Stall a Performance Presentation." **Computer**, June, p. 112.

[45] Kallfass, T. 1989. "Thin-Film Transistors for Addressing LC-Flat Panel Displays." **IEEE Proceedings, COMPEURO'89: VLSI and Computer Peripherals**, pp. 2-(20-23).

[46] Kendrick, D., A. Meeraus, and J. Alatorre. 1984. **The Planning of Investment Programs in the Steel Industry**. Published for the World Bank by the Johns Hopkins University Press, Baltimore and London.

[47] Kendrick, D.A., and A.J. Stoutjesdijk. 1978. **The Planning of Industrial Investment Programs: A Methodology**. Published for the World Bank by the Johns Hopkins University Press, Baltimore and London.

[48] Kernighan, B.W., and S. Lin. 1970. "An Efficient Heuristic Procedure for Partitioning Graphs." **The Bell System Technical Journal**, February, pp. 291–307.

[49] Kötzle, G. 1989. "VLSI Technology Trends." **IEEE Proceedings, COMPEURO'89: VLSI and Computer Peripherals**, pp. 5-(58–62).

[50] Kryder, M.H. 1989. "Data Storage in 2000-Trends in Data Storage Technologies." **IEEE Transactions on Magnetics**, November, pp. 4358–4363.

[51] Liebmann, W.K. 1989. "VLSI-the Driving Force for Computer Peripherals." **IEEE Proceedings, COMPEURO'89: VLSI and Computer Peripherals**, pp. P-(16–17).

[52] Mahon, M.J., R.B. Lee, T.C. Miller, et al. 1986. "Hewlett-Packard Precision Architecture: The Processor." **Hewlett-Packard Journal**, August, pp. 4–21.

[53] Markoff, J. 1991. "Supercomputing's Speed Quest." **The New York Times**, May 31.

[54] Marston, A., et al. 1987. "A 32b CMOS Single-Chip RISC Type Processor." **IEEE International Solid-State Circuits Conference: Digest of Technical Papers**, pp. 28–29.

[55] Martin, A. 1988. "Ruggedized Color CRT Assemblies." **Information Display**, September, pp. 10–13.

[56] Masterman, H.C. 1988. "Displays for Workstations." **Information Display**, November, pp. 11–14.

[57] McHaney, R. 1991. **Computer Simulation: A Practical Perspective**. San Diego, CA: Academic Press, Inc.

[58] Mee, C.D., and E.D. Daniel. 1990. **Magnetic Recording Handbook**. New-York: McGraw-Hill Book Company.

[59] " 'Micros' Vs. Supercomputers." 1991. **The New York Times**, May 6.

[60] Mills, R. 1989. "Why 3D Graphics?." **Computer-Aided Engineering**, March, pp. 50–68.

[61] Money, S.A. 1982. **Microprocessor Data Book**. New York: McGraw-Hill Book Company.

[62] "Motorola's 68K Series: Die Sizes." 1991. Provided by Roy Druian, **Motorola — 68000 Marketing and Applications Division**.

[63] "Motorola DRAM Data." 1991. Provided by Judy Racino, **Motorola - Marketing Communications Division**.

[64] Myers, W. 1991. "Five Plenary Addresses Highlight Compcon Spring 91: GaAs Targets 100-MHz-plus Computers." **Computer**, May, pp. 102–104.

[65] Myers, W. 1991. "The Drive to the Year 2000." **IEEE Micro**, February, pp. 10–13, 68–74.

[66] Nakanishi, H., S. Okuda, T. Yoshida, and T. Sugahara. 1986. "A High Resolution Color CRT for CAD/CAM Use." **Proceedings of the Society for Information Display**, $\underline{27}$(2), pp. 153–156.

[67] Ohsaki, T. 1991. "Electronic Packaging in the 1990's — A Perspective From Asia." **IEEE Transactions on Components, Hybrids, and Manufacturing Technology**, June, pp. 254–261.

[68] Opie, R. 1987. "Producing Color Graphic Display." **Control and Instrumentation**, October, pp. 47–48.

[69] "Panel Gives Chip Outlook." 1991. **The New York Times**, May 16.

[70] **PC Laptop Computers Magazine**. 1992. January.

[71] Peterson, J.L., and A. Silberschatz. 1987. **Operating System Concepts**. Reading, MA: Addison-Wesley Publishing Company.

[72] Phone conversations with Charles Malear, **Motorola — Advanced Microcontroller Division**, 1991.

[73] Phone conversations with Roger McKee, **Uniforum — Vice President of Marketing and Member Services**, 1991.

[74] Reichl, H. 1989. "Packaging of VLSI Devices." **IEEE Proceedings, COMPEURO'89: VLSI and Computer Peripherals**, pp. 5-(63-67).

[75] Riezenman, M.J. 1991. "Wanlass's CMOS Circuit." **IEEE Spectrum**, May, p. 44.

[76] Rosch, W.L. 1991. "Mainstream Monitors: What Really Matters." **PC Magazine**, July, pp. 103-186.

[77] Rosen, B., and S. Kriz. 1988. "Case Study: Developing a 3000-Line Interactive CRT Display." **Information Display**, January, pp. 12-15.

[78] Salus, P.H. 1991. "UNIX Software Next...." **COMPCON Spring '91: Digest of Papers**, pp. 362-364.

[79] Schadt, M. 1989. "Electro-Optical Effects, Liquid Crystals and their Application in Displays." **IEEE Proceedings, COMPEURO'89: VLSI and Computer Peripherals**, pp. 2-(15-19).

[80] Seidman, A.H., and I. Flores. 1984. <u>The Handbook of Computers and Computing</u>. New York: Van Nostrand Reinhold Company, Inc.

[81] Shmulovich, J. 1989. "Advanced Technology: Thin Film CRT Phosphors." **Information Display**, March, pp. 17-19.

[82] Stewart, G.A. 1988. "Multiscan Color Monitors." **Byte**, February, pp. 101-115.

[83] Stone, H.S., and J. Cocke. 1991. "Computer Architecture in the 1990s." **Computer**, September, pp. 30-38.

[84] Strum, W.E. 1988. "Trends in Microcomputer Image Processing." **SPIE Proceedings, 900, Imaging Applications in the Work World**, pp. 3-6.

[85] "Sun Workstations Pricing History and Performance Data." 1989. Provided by David Cohn, **Sun Microsystems, Inc. — District Sales Support Division**.

[86] "Sun Leads in Workstations." 1990. **The New York Times**, January 22.

[87] "Sun Workstations Performance Data." 1991. Provided by Andrea Pusateri, **Sun Microsystems, Inc. — Sales and Marketing Division**.

[88] Suntola, T. 1989. "Thin-Film EL-Displays." **IEEE Proceedings, COMPEURO'89: VLSI and Computer Peripherals**, pp. 2-(32–35).

[89] Sze, S.M. 1983. **VLSI Technology**. New York: McGraw-Hill Book Company.

[90] Takata, H. 1989. "Future Trend of Storage Systems." **IEEE Proceedings, COMPEURO'89: VLSI and Computer Peripherals**, pp. 1-(7–11).

[91] Tanksalvala, D., et al. 1990. "A 90MHz CMOS RISC CPU Designed for Sustained Performance." **IEEE International Solid-State Circuits Conference: Digest of Technical Papers**, pp. 52–53.

[92] Tannas, L.E., Jr. 1985. **Flat Panel Displays and CRTs**. New York: Van Nostrand Reinhold Company, Inc.

[93] Tannas, L.E., Jr. 1989. "Flat Panel Displays Displace Large, Heavy, Power-Hungry CRTs." **IEEE Spectrum**, September, pp. 35–36.

[94] Tasch, A. 1990. **Class Notes for EE396K, MOS-IC Process Integration**. Austin: Department of Electrical and Computer Engineering, The University of Texas at Austin.

[95] Terry, C. 1987. "Refinements in CRT Design Boost Resolution of Color Video Monitors." **EDN**, October 29, pp. 81–84.

[96] "The Open Software Foundation: A Look at Computing in the 1990s." 1991. **COMPCON Spring '91: Digest of Papers**, pp. 369–374.

[97] Tummala, R.R. 1991. "Electronic Packaging in the 1990's — A Perspective from America." **IEEE Transactions on Components, Hybrids, and Manufacturing Technology**, June, pp. 262–271.

[98] Virgin, L. 1987. "Understanding and Evaluating a Computer Graphics Display." **Information Display**, December, pp. 17–19.

[99] Waldecker, D. 1991. Presentation at the University of Texas at Austin: "IBM RISC System/6000."

[100] Weicker, R.P. 1990. "An Overview of Common Benchmarks." **Computer**, December, pp. 65–75.

[101] Wessely, H., O. Fritz, M. Horn, et al. 1991. "Electronic Packaging in the 1990's — A Perspective from Europe." **IEEE Transactions on Components, Hybrids, and Manufacturing Technology**, June, pp. 272–284.

[102] Wilczynski, J. 1989. "Low Temperature CMOS VLSI Technologies." **IEEE Proceedings, COMPEURO'89: VLSI and Computer Peripherals**, pp. 5-(73–78).

[103] Wood, R. 1990. "Magnetic Megabits." **IEEE Spectrum**, May, $\underline{5}$, pp. 32–38.

[104] Wurtz, J.E. 1989. "The Not-So-Amazing Survival of the CRT." **Information Display**, September, pp. 5–6, 18.

[105] Yetter, J., M. Forsyth, W. Jaffe, et al. 1987. "A 15MIPS 32b Microprocessor." **IEEE International Solid-State Circuits Conference: Digest of Technical Papers**, pp. 26–27.

[106] Yoffie, D.B., and A.G. Wint. 1987. "The Global Semiconductor Industry, 1987." **Harvard Business School, Case #9-388-052**.

[107] Zeidler, H.C. 1989. "Intelligent Access to Mass Memories." **IEEE Proceedings, COMPEURO'89: VLSI and Computer Peripherals**, pp. 1-(27–31).

Index

application program, 11
application specific IC, see ASIC
ASIC, 20, 62
assembled computer product
    dynamics of, 2
assembled workstation, 5
    component
        supply of, 4
    supply of, 4
assembly node, 133
attribute, 2, 58, 187
    state of, 57
average die testing time, 89

Bell Telephone Laboratories, 20
benchmark program, 17
    SPEC, 17
Berkeley UNIX system, 50
Bertsekas, D.P., 54
BiCMOS, 23
bipolar junction transistor, see
    BJT
bit, 9, 28
    density, 33, 82, 104
    per pixel, 45, 119
    per track, 31
    price per, 28
BJT, 20, 21
byte
    cost per, 11

    price per, 77

cache memory, 9, 11
    static random access memory, see SRAM
CISC, 14–17, 62
    instruction, 14
    microprocessor, 61
CISC CPU
    data, 70
        Intel, 70
        Motorola, 70
    die area, 64
    Intel, 62, 70
    IPC, 73
    MIPS model, 80
    Motorola, 62
clock
    cycle time, 16
    speed, 16
CMOS, see MOSFET
compiler, 14
    object code of, 15
    source code of, 15
component
    assembly, 11
    attribute, 57
    cost vs. single-unit price vs.
        bulk price, 143
    supply

dynamics of, 7
supply model, 2
  color cathode-ray tube display (CRT), 2
  integrated circuits (ICs), 2
  magnetic hard disk storage, 2
  UNIX operating system, 2
technology, 9
computer
  cycle time, 14
  industry, 2, 4, 5, 8
    dynamics of, 4, 187
    technology, 7
  initialization procedures, 11
  lifetime, 4
  period to obsolescence, 4
  simulation
    model, 56
    tool, 56
  startup, 11
  storage system, 27
  system, 2, 7, 9, 12
    attribute, 2, 8
    market niche, 2
    reliability, 26
    signal propagation, 18
    throughput, 15, 67
  task
    execution, 9
    interaction with the I/O interface, 9
  technology, 7
  use style, 2
  user environment, 2
  workstation, 2, 4
computer/user window-interface manager, 9
computing environment

distributed, 8
  DEC and Xerox, 8
multitasking
  DEC and Xerox, 8
contributions, 4
CPU, 5, 20, 37
  architecture, 70
  CISC, see CISC CPU
  instruction, 11
    pipelining, 15
  price per megaHertz, 6
  RISC, see RISC CPU
  speed, 60, 70
    model, 80
  throughput, 70
  wait-on-data cycle, 11
Crookes, W., 37
CRT, 2, 5, 9, 12, 37, 38, 42, 44, 109, 187
  aspect ratio, 40
  bandwidth, 45, 116, 119
  bulb
    faceplate, 40
    funnel, 41
    neck, 40
  cathode ray, 40
  cathodoluminescence phenomen
    42
  color phosphor dots layout, 41
  color sources
    red, green, and blue (RGB), 42
  cost model, 121
  data, 112
  display component, 38
  hardware driver, 46, 48, 117
    buffer, 46
    controller, 46

supervisory processor, 46
video generation system, 46
history, 37
image driver, 12
image generating technique, 44
interlaced scanning, 44
maximal resolution, 47
maximum resolution, 117
metal shadow mask, 42, 47
  expansion, 48
  hole, 42
  hole pitch, 44, 47, 112
  manufacturing yield, 120
  model results vs. actual market data, 159
noninterlaced scanning, 44
resolution, 46, 110, 117
  lithography, 112
supply model, 115

DARPA, 50
data
  reliability, 28, 33, 36
  storage space, 8
  storage system, 8n, 12
    access rate, 31
    channel, 31
    capacity, 2
    data rate, 2
    magnetic, 8n
    optical, 8n
  transfer, 35
    rate, 36
DEC, 13
  Ultrix, 54
deterministic
  modeling methodology, 56
  simulation model, 2

die, 18
  area, 25, 64, 70
    fixed capability IC, 83
  configuration, 65
  cost, 18
  manufacturing yield, 64, 86
  operational speed, 25, 64
  size, 62
digital
  data bit, 28
  information, 9n, 26
  magnetic recording, 91
direct access storage device (DASD), 28
direct cost, 72
display, 8
  capability, 8
  color template, 12
  CRT, 38
  electroluminescent, 38
  hardware, 37
    driver, 110
  liquid crystal, 38
  pixel, 8
  plasma, 38
  refresh rate, 8, 12
  resolution, 2, 8, 12
  technology, 37
    Japan, 38
distributed computing environment, 2, 5
DRAM, 2, 5, 11, 12, 15, 20, 25, 33, 37, 46, 48, 61, 62, 187
  access speed, 16
  capacity, 2
    model, 82
  data, 77
  density, 2

die area, 64
Japanese competition, 77
Motorola, 77
nonvolatile, 156, 177
price per megabyte, 5, 6
dynamic random access memory,
    see DRAM

EEPROM, 11n
electric field controlled current, 21
electron, 19, 21
    beam, 41, 47
        accelerating aperture, 41
        deflection aperture, 41
        focus aperture, 41
        misconvergence, 48
    gun, 44
    switching speed, 47
Ethernet hardware, 8

fault tolerance, 26
FET, 21
    JFET, 21
    MOSFET, see MOSFET
field-effect transistor, see FET
fixed capabilities workstation, 5
floating point
    instruction, 14
    processor, 15
    unit, see FPU
FPU, 11, 16, 67

GaAs IC, 24
gallium arsenide (GaAs), 19, 24
General Purpose Simulation System (GPSS), 56
germanium (Ge), 19
Ghausi, M.S., 20, 22, 25
Gordon, Geoffrey, 56

graphics microprocessor
    CISC, 119

Hamacher, V.C., 12
hardware
    bus, 9
    capability, 5
    communication, 12
        delay, 12
        line, 9
    component, 8
    assembly, 9
Hennessy, J.L., 12, 64, 86, 146n
    MIPS project, 14
high definition display, 12

I/O interface, 9, 11
    external, 9
IBM, 14
    801 minicomputer project, 14
    AIX, 54
    DASD, 28, 92
    magnetic hard disk
        head, 34
    RISC System/6000 workstation, 16
IC, 2, 18, 20, 21, 26, 61
    burn-in cost, 90
    circuit configuration
        doping, 67
        etching, 67
    CISC CPU
        model results vs. actual
            market data, 147
    cost model, 84
    critical mask, 62, 67
    data, 70
    die area, 62
    die yield

INDEX

model results vs. actual market data, 146
direct cost, 72
DRAM
  model results vs. actual market data, 151
etching process, 25
fabrication laboratory, 61
feature size, 5, 25, 26, 62
  AT&T, 26
  IBM, 26
  physical barrier, 169
  trend, 64
final test yield, 90
functionality, 18
hybrid, 21
Japanese fabrication laboratories, 62
layout, 26
list price, 70
manufacturing, 62
  capital intensity, 61
  clean room technology, 181
  cost, 84
  environment, 6
  learning, 181
  masking levels, 62
  processing steps, 62
model results vs. actual market data, 146
monolithic, 21
packaging
  cost, 89
  technology, 67
production trend, 62
reliability, 6, 62
RISC CPU
  model results vs. actual market data, 149

superconductive, 24
supply model, 78
  period of study, 78
  testing cost, 87
integer unit, *see* IU
integrated circuit, *see* IC
IU, 11, 16, 67

key findings, 5

LAN, 13, 50
  software, 8
laser diode, 24
LCD, 9, 12, 38, 109
liquid crystal display, *see* LCD
list price, 72
lithography, 25, 65
  optical, 25
local area network, *see* LAN

magnet, 27
  polarization, 27
magnetic bit cell, 28, 30, 33, 91
  layout, 35
  length, 35, 94, 156
magnetic byte, 30
magnetic hard disk, 2, 5, 11, 187
  actuator, 95
    electromagnetic, 34, 35
    positioning time, 35
  areal density, 12, 98, 106
  component, 28
  cost model, 106
  data, 96
    transfer time, 35
  data channel, 96
  data rate, 12, 104
    note on, 159
  data request, 34
  drive, 28

head-actuator setup, 95
head-positioning servomechanism, 34
inductive magnetic head, 32
　reading process, 33
　technology-driving trend, 94
　writing process, 33
model results vs. actual market data, 156
performance, 34
price per megabyte, 96
reliability, 92
response time, 34, 35
servo, 34, 95
　disk, 34
servomechanism, 34
storage technology, 27
supply model, 100
technology-driving trend, 94
track, 30, 35
　density, 31, 36, 104
　pitch, 30
　sector, 30
volumetric density, 106
note on, 158
magnetic medium, 30
　layout, 30
　technology-driving trend, 94
magnetic recording
　IBM, 91
magnetic storage, 5, 27
　history, 28
　system, 28, 92
　technology, 91
main memory, 8, 11, 12
　board, 11, 27
mainframe, 4, 15n, 91, 139
　HP

　　RISC Precision Architecture, 168
　IBM
　　superscalar RISC architecture, 168
　market share, 151
　technology, 151
manufacturer, 143
　CRT, 109, 110, 112, 121, 161
　Japan, 49
　IC, 23–25, 72, 77, 86, 87, 90, 147, 149, 153, 177, 188, 189
　magnetic hard disk, 98, 104, 156, 179
　mainframe, 168
　silicon wafer, 86
　supercomputer, 24, 168
　workstation, 125, 133, 161, 190
mathematical modeling, 7
Mee, C.D., 31, 34
memory, 61
　board, 9
　caching, 15, 16
　chip, 8
　management unit, see MMU
metal-in-gap (M-I-G) ferrite head, 32, 33
MFLOPS, 17
microcontroller, 8, 61
micromechatronics, 35, 95
microprocessor, 2, 8, 11, 18, 61, 187
　architectural configuration, 9
Microsoft
　Xenix, 54
microwave circuits fabrication
　GaAs, 24

MIPS, 17, 18, 70, 77, 78, 96, 108, 109, 119, 121, 131, 143, 147–149, 151
   metric, 18
MMU, 11, 16, 67
MOSFET, 21
   CMOS, 20, 22, 67
   speed, 22
   NMOS, 21, 22
   PMOS, 21
multitasking machine, 12n

NMOS, *see* MOSFET
number of clock cycles per instruction, 15
number of CPU instructions per cycle, *see* CISC/RISC CPU IPC
number of semiconductor wafer defects, 18

obsolescence, 65
   rate, 61
Ohm's law, 17
Open Software Foundation (OSF), 124
open system software, 49
operating system, 8, 27, 49
   development-from-scratch cost, 49
   distributed computing, 50
   multitasking, 50
   networking, 50
      capability, 8
   porting, 52
   cost, 49
   system resources manager, 9
optical
   computer, 24

   technology
      photonic transmission of signals, 24

parallel computer, 16
   MIMD, 16n
   SIMD, 16n
Patterson, D.A.
   RISC 1 project, 14
permanent storage component, 5
personal computer (PC), 2, 8
Peterson, J.L., 52, 54
phosphor
   depositing process, 44
   dot, 48
   RGB triad, 42
photoresist, 25, 65
   mask, 67
   process, 42
pins per
   bipolar package, 67
   CMOS package, 67
pixel, 44, 45, 110
Posix, 54
process model, 133, 134, 139
processing
   board, 9, 11, 27
      assembly, 9
      interconnection delay, 67
   power, 2, 8
      functionality, 2n
      speed, 2n
   stage, 133
   unit
      instruction, 11
processor, 11, 12, 15, 16, 62
   assembly language, 73
   branch instruction, 73
product

life, 61
yield, 61
productive unit, 133
program
   execution time, 13
   number of instructions, 14

read-only-memory, see ROM
real-time
   application, 189
   computing, 5, 189
relational diagram, 57, 58
   concept, 187
   CRT
      cost per megapixel, 121
      hardware driver, 117
      example, 60
   IC
      average testing time, 88
      DRAM capacity, 82
      operational speed, 80
   magnetic hard disk
      cost per megabyte, 106
   UNIX
      development-from-scratch time, 128
      porting time, 129
   workstation
      hardware attributes, 131
resolution, see CRT resolution
RISC, 14, 16, 17
   compiler, 15
   memory access instruction, 15
   microprocessor, 61
RISC CPU
   data, 70
   Hewlett-Packard Precision Architecture (HP-PA), 73
   IPC, 73
   MIPS model, 80
   Sun Microsystems Scalable Processor Architecture (SUN-SPARC), 73
Ritchie, Dennis, 50
ROM, 11

Sematech consortium, 177
semiconductor
   n-type, 19
   pn-junction, 20, 21
   p-type, 19
   DRAM, 5
      nonvolatile, 5
   extrinsic, 19
   fabrication technology
      bipolar, 20
      CMOS, 61
      NMOS, 21
      PMOS, 21
   factory, 61
   hole, 19, 21
   industry, 8, 20, 61
   ingot, 18
   intrinsic, 19
   resistivity, 19
   substrate, 25
   transistor, 20
   wafer, 18
semiconductor DRAM vs. magnetic hard disk, 154, 179, 188
sensitivity analysis
   feature size, 169
   19-inch color CRT price, 174
   assembled workstation price, 177
   CPU price per megaHertz, 170

DRAM price per megabyte, 172
magnetic hard disk price per megabyte, 174
price per megabyte: DRAM vs. magnetic storage, 177
number of silicon wafer defects per unit area
CPU price per megaHertz, 183
die yield, 182
DRAM price per megabyte, 183
technology-driving trend change in, 168
Shockley, W.B., 20
silicon, 19
  crystal lattice, 19
  ionization, 19
  thin film package, 67
  wafer
    cost, 85
    defect, 6, 188
    diameter, 62, 68
simulation, 57
  continuous, 56
  discrete
    data, 58
    event, 56
  modeling approach, 55
  Monte Carlo, 56
software
  component, 8, 9
  assembly, 9
Sony
  Trinitron, 112
SRAM, 11, 23
storage technology, 9
  magnetic, 9

nonvolatile, 27
optical, 9
suggestions for future research, 189
Sun Microsystems, 50
  SunOS, 54
supercomputer, 4, 24, 24n, 167
  Cray Research, 167n
superconductivity, 24
supply model
  discrete event simulation, 141, 168, 187
  dynamic behavior, 141
  simulation
    code, 142
    input parameter, 143
    tuning, 142
    utility, 141
system hardware capabilities, 8

Tannas, L.E. Jr., 41, 44
technological attribute, 57
technology-driving trend, 4, 5, 57–59, 68, 187
TFM head, 32
  photolithography, 32
thin film metal head, *see* TFM head
Thompson, Ken, 50

UNIX, 2, 12, 13, 49, 50, 124, 187
  AT&T, 54
  computer security, 53
  data, 124
  deadlock handling, 53
  development-from-scratch
    cost model, 132
    data, 125
    supply model, 126

distributed computing, 53
distributed environment, 13
history, 50
kernel, 50, 52
memory management, 53
model results vs. actual market data, 161
multitasking, 53
network communication, 53
porting
  cost model, 132
  data, 125, 126
  supply model, 126
shell, 50
software attribute, 130
system
  call, 51
  program, 50
user
  application, 8
  interface, 8

VAX 11/780, 17
very large scale integration, *see* VLSI
video adapter, *see* CRT hardware driver
VLSI, 36, 48, 110

Wanlass, Frank, 22
wave reflection phenomenon, 17
Welker, Henry, 24
wide area network (WAN), 13
workstation, 2, 5, 7–9, 11, 13, 16, 187
  advanced PC, 8
  assembly network, 134
  assembly plant, 133
  assembly process, 134

assembly process model, 55, 187
  constraints, 138
  input parameters, 164
  intermediate product, 134
  objective function, 137
  projected results, 167
assembly steps, 136
birth of, 8
component
  assembly, 7
  attribute, 142
  supply model, 5
display, 110
environment, 13
hardware attribute, 130
performance, 8, 17
price ceiling, 5
supply
  dynamics of, 7
workstation attribute, 13, 164
  display
    color, 13
    resolution, 13
  DRAM
    amount of, 13
  operating system
    multitasking capabilities, 13
    networking capabilities, 13
  performance, 8, 13
  price, 8
  processing board
    speed, 13
  storage system
    capacity, 13
    response time, 13
    total cost, 13
workstation component supply

dynamics of, 56
supply model, 55

X windows manager, 13